Tobias Kennerknecht

Fatigue of Micro Molded Materials – Aluminum Bronze and Yttria Stabilized Zirconia

Schriftenreihe
des Instituts für Angewandte Materialien
Band 45

Karlsruher Institut für Technologie (KIT)
Institut für Angewandte Materialien (IAM)

Eine Übersicht aller bisher in dieser Schriftenreihe erschienenen Bände finden Sie am Ende des Buches.

Fatigue of Micro Molded Materials – Aluminum Bronze and Yttria Stabilized Zirconia

by
Tobias Kennerknecht

KIT Scientific Publishing

Dissertation, Karlsruher Institut für Technologie (KIT)
Fakultät für Maschinenbau
Tag der mündlichen Prüfung: 25. März 2014

Impressum

KIT Scientific Publishing

Karlsruher Institut für Technologie (KIT)
KIT Scientific Publishing
Straße am Forum 2
D-76131 Karlsruhe

www.ksp.kit.edu

Print on Demand 2014

ISSN 2192-9963
ISBN 978-3-7315-0293-7
DOI 10.5445/KSP/1000043832

Fatigue of Micro Molded Materials – Aluminum Bronze and Yttria Stabilized Zirconia

Zur Erlangung des akademischen Grades
Doktor der Ingenieurwissenschaften
der Fakultät für Maschinenbau
Karlsruher Institut für Technologie (KIT)

genehmigte
Dissertation
von

Dipl.-Ing. Tobias Kennerknecht

Tag der mündlichen Prüfung:	25.03.2014
Hauptreferent:	Prof. Dr. rer. nat. Oliver Kraft
Korreferent:	Prof. Dr.-Ing. Martin Heilmaier

Danksagung

Die vorliegende Arbeit entstand am Karlsruher Institut für Technologie (KIT) in der Nachwuchsgruppe des Sonderforschungsbereiches SFB499 zur Entwicklung, Produktion und Qualitätssicherung urgeformter Mikrobauteile aus metallischen und keramischen Werkstoffen. Diese Nachwuchsgruppe war am Institut für Angewandte Materialien (Werkstoff und Biomechanik) des KIT beheimatet und wurde, wie auch meine Arbeit, dankenswerter Weise von der Deutschen Forschungsgemeinschaft (DFG) finanziert. Für weitere finanzielle Unterstützung möchte ich mich beim Karlsruhe House of Young Scientists des KIT bedanken, durch welche mir ein zweimonatiger Auslandsaufenthalt an der Johns Hopkins Universität in Baltimore ermöglicht wurde.

Herzlich bedanken möchte ich mich bei Prof. Oliver Kraft für die Übernahme des Hauptreferats, für seine Anregungen und Diskussionsbeiträge, sowie für die stetige Unterstützung der Nachwuchsgruppe. Prof. Martin Heilmaier danke ich für die Bereitschaft das Korreferat zu übernehmen.

Dr. Christoph Eberl danke ich besonders für die Betreuung meiner Arbeit und die Möglichkeit diese in der von ihm geleiteten Nachwuchsgruppe durchzuführen. Seine ständige Diskussionsbereitschaft, zahlreichen Anregungen und seine motivierende Begeisterung für die Wissenschaft haben maßgeblich zum Gelingen der Arbeit beigetragen. Auch für den bereichernden Kontakt zu zahlreichen internationalen Wissenschaftlern, den er und Prof. Oliver Kraft mir im Rahmen von Konferenzbesuchen und Kooperationen ermöglicht haben, möchte ich mich herzlich bedanken.

An vorderster Stelle ist hier mein Auslandsaufenthalt an der Johns Hopkins Universität in Baltimore zu nennen. Diesbezüglich danke ich Prof. William N. Sharpe, Jr. und Prof. Kevin Hemker, dass sie mich an Ihrem Institut als Gast aufgenommen haben. Sehr lehrreich und angenehm war die Arbeit im Labor mit Prof. William N. Sharpe, Jr. und Prof. Chung-Youb Kim.

Ebenso danke ich für die regelmäßigen Besuche von Prof. John Balk und die gemeinsame Durchführung von Experimenten an nanoporösem Gold in sehr angenehmer Atmosphäre sowie mit stets spannenden Herausforderungen und Ergebnissen.

Für zahlreiche Anregungen und Diskussionen bezüglich der Untersuchungen an Zirkonoxid möchte ich Dr. Martin Härtelt herzlich danken, der stets bereit war mit seinem Rat zur Verfügung zu stehen. Außerdem danke ich Dr. Theo Fett und Prof. Peter Gumbsch für eine Diskussion der Ergebnisse.

Fatih Çetinel danke ich für die Bereitstellung seiner Bruchfestigkeitsdaten der Zirkonoxid-Proben, sowie für die Probenherstellung. Ebenso danke ich Durime Buqezi-Ahmeti und allen daran Beteiligten des IAM-WPT am KIT für die Probenherstellung. Auch den übrigen Kollegen des SFB499 danke ich für die kollegiale Zusammenarbeit.

Für die Hilfe am Dual-Beam-Mikroskop danke ich Daniela Exner, insbesondere für die REM-Bilder der Zirkonoxid-Proben. Für Untersuchungen mittels EDX, EELS, TEM und STEM danke ich den Mitarbeitern vom LEM des KIT, insbesondere Herrn PD Dr. habil. Reinhard Schneider für die gemeinsamen und lehrreichen Sitzungen am TEM. Auch Dr. Reiner Mönig, Dr. Dominik Kramer und Dr. Matthias Funk danke ich für die Unterstützung bei EDX-Untersuchungen.

Ferner danke ich für die Unterstützung bei den Experimenten an Zirkonoxid durch die Studienarbeit von Sandy Pelletier, bei Finite Elemente Simulationen durch Thomas Straub und Geoffroy Bretzner, der auch bei CAD-Konstruktionen geholfen hat. Für weitere studentische Hilfe danke ich Benjamin Hertwick, Anson Santoso Wong und Thomas Ward.

Ganz besonders möchte ich mich bei Ewald Ernst für die Herstellung von Versuchsstandkomponenten und die außerordentliche Hilfsbereitschaft danken. Außerdem danke ich allen Institutskollegen des IAM-WBM für die äußerst kollegiale Atmosphäre und die gegenseitige Unterstützung. Bei Matthiew Berwind möchte ich mich für die Sprachkorrekturen bedanken.

Herzlicher Dank gebührt auch meiner Familie, meinen Freunden und Diana, die mich trotz des zum Entstehen dieser Arbeit nötigen Verzichts auf gemeinsame Zeit stets unterstützt haben.

Tobias Kennerknecht

Fatigue of Micro Molded Materials – Aluminum Bronze and Yttria Stabilized Zirconia

255 Seiten, 89 Abbildungen, 19 Tabellen

Kurzzusammenfassung

Um Ermüdungseigenschaften von Mikroproben (mit Breiten und Dicken in der Größenordnung von 100 µm bis 200 µm) testen zu können, müssen üblicherweise selbstentwickelte Versuchsaufbauten verwendet werden, da geeignete kommerzielle Apparaturen kaum verfügbar sind. Daher wurde eine flexible Mikro-Prüfmaschine entwickelt, mit der quasistatische sowie zyklische Zug-, Druck- und Biegeexperimente bis 100 Hz durchgeführt werden können. Die Hauptkomponenten, welche eine zuverlässige kraftkontrollierte Ermüdung der Mikroproben ermöglichen, sind eine Datenerfassungskarte mit FPGA Technologie (Field Programmable Gate Array für Signalmessung und -ausgabe), ein Piezoaktor und eine dynamische Kraftmesszelle. Für höhere Frequenzen wurde eine piezogesteuerte, in Resonanz arbeitende Prüfapparatur entwickelt, welche die erste Zug-Druck-Eigenform der Probe bei bis zu 2 kHz anregt. Die Kraft wird über die kapazitiv gemessene Verschiebung einer Masse errechnet und geregelt, welche an der Probe angebracht ist.

Mikrogegossene Proben aus Aluminiumbronze (CuAl10Ni5Fe4) wurden hauptsächlich im HCF- (high cycle fatigue) und VHCF-Bereich (very high cycle fatigue) ermüdet. Untersuchungen mit Rasterelektronenmikroskop, Ionenstrahlmikroskop und Transmissionselektronenmikroskop zeigten, dass die Mikrostruktur das Rissfortschrittsverhalten stark beeinflusst. Die Ermüdungsrisse verlaufen bevorzugt entlang 200 nm dicker lamellarer Ausscheidungen, welche ungefähr 45° zur aufgebrachten Zuglast orientiert sind. Treffen sie auf runde Ausscheidungen (mit Durchmessern von wenigen µm), werden sie um diese herum abgelenkt. Lange Ermüdungsrisse, welche bei niedrigen Lastamplituden entstanden, wiesen häufig einen mäandernden Risspfad auf. Dieser enthielt stellenweise gerade, in Richtung maximaler Scherspannungen verlaufende Abschnitte. In einigen Fällen wurden an mehreren Stellen der Probenoberfläche Risse sichtbar, welche mehrheitlich ähnlich orientiert waren. Obwohl die untersuchten Proben aus mehreren Herstellungschargen stammten und verschiedene Frequenzen untersucht wurden, sind die Ergebnisse relativ homogen und ähnlich zu Werten makroskopischer Proben (kein Ermüdungsbruch unterhalb einer Amplitude von 190 MPa). Dies ist auf den Fertigungsprozess des Mikrogießens zurückzuführen, bei dem nicht nur die Probengeometrie, sondern auch die Mikrostruktur auf eine kleinere Größe skaliert wird.

Einige Proben wurden auf Vorder- und Rückseite poliert, was keinen Einfluss auf die Lebensdauer hatte, jedoch die Oberflächenanalyse nach dem Versuch erleichterte. Dabei wurden verschiedene Merkmale gefunden, die wahrscheinlich auf Oxidbildung zurückzuführen sind. Ferner wurde aus den Rissen hervortretendes Material beobachtet, was großen Extrusionen ähnelt.

Mikrogegossene Biegebalken aus Zirkonoxid (3Y-TZP) wurden unter Dreipunktbiegung zyklisch belastet. Die Ergebnisse wurden unter Anwendung eines statistischen Verfahrens mit Festigkeitsdaten des Probenherstellers kombiniert. Somit konnten Risswachstumskurven errechnet werden. Die Ergebnisse zeigten, dass ein Mindestmaß an Rissfortschritt nötig ist, um eine zyklisch degradierbare Abschirmung der Rissspitze zu entwickeln. Hochfeste Probenchargen mit sehr kleinen Rissverlängerungen bis zum Bruch zeigten einen Risswachstumsexponenten von 31, was verglichen mit Literaturwerten für 3Y-TZP im Bereich von reinem unterkritischen Risswachstum liegt.

Probenchargen, welche hingegen eine niedrigere Festigkeit und somit eine größere Rissverlängerung bis zum Bruch aufweisen, hatten einen Risswachstumskoeffizienten von 22. Dies stimmt gut mit Literaturwerten für zyklisch belastetes 3Y-TZP überein. Lastratenabhängige Bestimmungen der Biegefestigkeit deuteten darauf hin, dass bei Mikroproben sehr viel höhere Lastraten aufgebracht werden sollten, als es für makroskopische Proben üblich ist, um den Einfluss von unterkritischem Risswachstum auszuschließen. Dieses Phänomen könnte auf Skalierungseffekte zurückzuführen sein. Es sollte jedoch mit weiteren Untersuchungen abgesichert werden, da diesbezüglich nur sehr wenige Experimente durchgeführt wurden.

Untersuchungen am Rasterelektronenmikroskop zeigten ähnliche Bruchbilder, wie sie von makroskopischen Proben bekannt sind. Einige Merkmale auf der Bruchfläche, wie zum Beispiel das typische Erscheinungsbild von Bruchauslösern, sind häufig jedoch nur schwer auf der kleinen Bruchfläche von Mikroproben (200 µm × 200 µm) zu identifizieren.

Tobias Kennerknecht

Fatigue of Micro Molded Materials - Aluminum Bronze and Yttria Stabilized Zirconia

255 pages, 89 figures, 19 tables

Abstract

Testing fatigue properties of micro samples (having a width and a thickness on the order of 100 µm to 200 µm) requires the use of custom built devices, since standardized commercial facilities are not available. Therefore, a flexible micro sample tester was developed, which allows for quasistatic tensile, compression and bending tests as well as for the corresponding cyclic investigations up to 100 Hz. A field programmable gate array data board, a piezo actuator and a dynamic load cell are the key components to ensure reliably load controlled cycling. For higher frequencies, a resonant piezo driven setup was developed, working up to 2 kHz in the first push-pull mode of the sample. The load is calculated and controlled using the capacitively measured displacement of a mass, which is attached at the sample.

Micro molded aluminum bronze samples (CuAl10Ni5Fe4) were mainly fatigued in the high and very high cycle regime. The microstructure dominates the crack propagation behavior, as it was shown by SEM, FIB and TEM analyses. Cracks propagate preferentially along lamellar precipitates, which are oriented at about 45° with respect to the tensile load and are about 200 nm thick. Round-shaped precipitates (few µm in diameter) are contoured by the crack. For longer near threshold cracks, a corrugated crack path was observed, sometimes containing straight sections along the direction of maximum applied shear. In some cases, cracks appeared at several locations on the sample, mainly oriented in similar directions. Despite the fact that several frequencies and samples coming from different molding batches were used, the experiments show quite uniform results, similar to macroscopic samples (no failure below 190 MPa amplitude). This is attributed to the manufacturing process, scaling down not only the dimensions, but also the microstructure of the samples.

Some samples were polished on the front and back sides. No impact on the lifetime could be found from polishing, but it helped surface analysis after fatigue. Several features were found, which are likely related to the formation of oxide. Material coming out of cracks was observed, which is similar to large extrusions.

Micro molded zirconia bars (3Y-TZP) were subjected to cyclic three-point-bending testing. The data was statistically combined with strength values from the sample manufacturer, in order to generate crack growth curves. The results gave evidence that a minimum crack extension is necessary to develop cyclically degradable shielding. High strength samples with a small crack extension until failure showed a crack growth exponent of 31, which is in the range of purely subcritical crack growth for 3Y-TZP reported in the literature. However, samples having a lower strength and thus a higher crack extension until failure showed a crack growth exponent of 22, which is the range reported in the literature for cyclically loaded 3Y-TZP samples. Rate dependent bending tests indicated that much higher load rates should be applied as it is common for macro samples, in order to prevent subcritical crack growth to occur. This effect might be attributed to scaling effects, but it should be verified in future studies, since only very few tests were conducted on this topic. SEM investigations showed similar failure morphologies, as they are reported for macroscopic samples. However, typical marks on the fracture surface, such as mirrors, known for macro samples are difficult to be identified on the small fracture surface (200 µm × 200 µm) of the investigated micro samples.

Contents

1. Introduction

> "I possess no direct experiments bearing on this point. But that the alloy has considerable elasticity is unquestionable. I may here state that an eminent Parisian instrument-maker informed me, that of all the wires tried for the suspension of Foucault's Pendulum for illustrating the rotation of the earth, none, not even those of steel, were so durable under that severe ordeal as wires made of aluminium bronze. It would appear, therefore, to be the most proper material for the suspension springs of clock pendulums."[1]

These words were published in 1863 by Lieutnant-Colonel A. Strange, who was instructed by his government to design high level instruments of physical research, mainly for geodesy and astronomy. The ability for mass production by casting combined with the transportability and the need for high quality of the products made Strange consider the novel material aluminum bronze as an appropriate alloy. In his investigations he found excellent results for this material compared to others, such as gun metal and brass, regarding the following aspects: Tensile strength (better than "Krupp's famous cast steel"), resistance to compression, malleability (excellent forging), transverse strength ("3 times more rigid than gun metal"), expansive ratio ("less than gun metal"), founding qualities ("admirable castings of any size"), behavior under files and cutting tools ("it leaves nothing to be desired"), resistance to atmospheric influences ("tarnishes much less [...] than [...] gun metal, brass, silver, cast iron, or steel"), fitness to receive graduation ("The lines are very distinct under the microscope"), elasticity (see citation above), fitness for being made into tubes ("It admits of every process necessary for this purpose"), specific gravity ("nearly the same as that of wrought iron, and less than that of either brass or gun metal").[1]

Nowadays, such advantageous properties would sometimes be described in other words, but the finding that aluminum bronze is an interesting candidate unifying excellent workability, very good mechanical properties and corrosion resistance is still valid. In the more recent and detailed review of aluminum bronze by Meigh, the attractive properties of this material are described with "high strength", "exceptional resistance to corrosion", "excellent resistance to cavitation", "castable by all the main processes", "pressure tight", "ductile and malleable", "weldable", "good machinability", "good shock resistance", "exceptional resistance to fatigue", "good damping", "suitable at high temperatures", "suitable at low temperatures", "good wear resistance", "low magnetic permeability", "non-sparking" and the "attractive appearance" [2] (pp. 3-4).

According to Strange, the international exhibition called his attention to aluminum bronze, which was a mixture of copper and aluminum elaborated by Dr. Percy in about 1857, showing the best mechanical properties for an aluminum content of 10 %. [1]

Later on, other alloying elements were added to develop the aluminum bronze. The most important ones, presented by Meigh in detail, are the binary system Cu-Al, the ternary systems Cu-Al-Fe, Cu-Al-Ni, Cu-Al-Mn, Cu-Al-Si, Cu-Al-Be, Cu-Al-Zn, Cu-Al-Co and last but not least the Cu-

Al-Ni-Fe system, sometimes also containing small portions of Mn, which is part of the investigations from the present work. [2]

The latter is mainly used for applications in harsh environments, such as tubings or machine parts for the chemical industry and for the food industry as well as for naval components such as propellers. This is again due to its high resistance to corrosion, cavitation, and wear, high strength and good resistance to fatigue, and its castability and weldability. Further applications are gear wheels, worm gears, bushings, joints and (nonsparking) tools for production processes.[3]

One major reason why the alloy is limited to quite special applications should be its elevated price. This criterion is of minor interest for micro parts (parts having a width and a thickness of several tens or several hundreds of micrometers), where low volumes of material are needed and the higher portion of the product price is related to the fabrication process. Thus, aluminum bronze is an interesting candidate to produce micro parts en masse by casting, without the necessity of subsequent process steps. The resulting parts would be mechanically and chemically resistant, which is an important point for novel miniaturized products, such as micro turbines, micro gears or micro tools. These kind of parts are appropriate to design micro robots enabling inspections or reparations of tiny tubing systems. For instance, technical systems such as water tubings, oil tubings etc. or even human systems such as blood vessels - in this case a biocompatible encapsulation would be indispensable, of course - might get accessible by means of such devices. Furthermore, miniaturized reactors for chemical analyses with small quantities of substances could be equipped with such micro-parts. Also complex robotic systems like roving vehicles or robots for explorations in water (or other liquids) or air might profit by an ongoing miniaturization of mechanically resistant and reliable parts. The more complex the robot itself or the explored system is, the more important the reliability of the device is, in order to prevent expensive damage.

Thus, special care has to be taken in judging the material behavior of micro-parts in terms of reliability. The observations made on macroscopic parts can differ remarkably from the reliability behavior of micro-parts. Scaling effects, such as a rising surface to volume ratio for parts, which are scaled down in size, might for instance lead to a higher impact of surface reactions like corrosion. At the same time, heat inertia is expected to be much smaller for micro-parts due to this scaling effect. Furthermore, the specific weight might lose importance, since the stiffness with respect to the weight of a part rises, when it is scaled down. When the diameter of a gear wheel is scaled down, the weight decreases quadratically with the diameter (not taking into account a reduction in thickness, which would induce an additional decrease of weight), which makes it easier to be accelerated and reduces the inertial loads at first view. However, when the transmitted power or the circumferential speed of a smaller wheel should be comparable to the one of a larger wheel, it must turn at a higher frequency (scaled up about the same ratio as the wheel diameter was scaled down). In this simplified analysis, important points like changes in terms of efficiency due to larger influences of friction or adhesion at the small scale are not considered. Anyhow, it becomes clear that the requirements for a material as well as the impact of physical phenomena on the reliability of a part change, when it is drastically scaled down.

The higher frequency, at which the smaller wheel turns in the given example, leads to a higher number of loading cycles, which is reached over time. This is typical for micro parts, since their lower weight allows higher actuation frequencies. Their natural frequencies are elevated due to their high stiffness to mass ratio. Small volumes or smaller loads applicable on a smaller surface must be compensated by high speeds in order to attain high power devices. Thus, the very high cycle fatigue range, where loading cycles of 1×10^8, 1×10^9 and more are applied, is of special interest regarding micro parts, and therefore was part of the present investigation.

A further aspect, which has particularly high relevance for micro-parts, is the interaction with the microstructure. Since the dimensions of the parts reach the order of the size of microstructural features such as grain sizes, sizes of precipitates or inclusions, the distance between precipitates or inclusions, etc., deformation mechanisms or failure mechanisms can differ in comparison to macroscopic parts. Furthermore, micro parts are usually manufactured using special processes, for which reason their microstructure can differ remarkably from macroscopic samples. Thus, there is a need to investigate micro samples rather than to extrapolate material properties found for macroscopic samples to the microscale. For this purpose, micro sample testing devices were developed in this work, enabling quasistatic and dynamic tensile tests, compression tests as well as bending tests.

The latter were performed on ceramics samples made out of yttria stabilized zirconia. This material is also of high technical interest. One reason is its elevated strength and plasticity, which lead to sensational titles such as "Ceramic steel?" published in the journal "Nature" in 1975 [4]. For this reason, it serves today as a high-end material e.g. for consumer coffee grinders, knife blades, and dental implants. Zirconia is also an interesting material for high temperature applications. The elevated plasticity contributes to thermal shock resistance. Typical technical applications are wire-drawing tools, crucibles, thermal insulation layers (also called thermal barrier coatings, TBC) or bearing components [5] (p. 6). Due to its ion conductivity at elevated temperatures, zirconia is also applied for high temperature fuel cells.

A wide range of different kinds of zirconia ceramics has to be distinguished. For all of them, the mechanically or thermally induced phase transformation from the tetragonal to the monoclinic phase plays a more or less important role, concerning the mechanically interesting properties.

In the present work, yttria stabilized zirconia was investigated, which belongs to the group of tetragonal zirconia polycrystals (TZP). TZP consist of fine grains, which are mainly tetragonal except of a small portion of cubic phase [6]. Second, there are dispersion toughened zirconia, where toughening zirconia particles are introduced in a ceramic matrix, such as alumina (zirconia toughened alumina, ZTA) or mullite (ZTM). Third, partially stabilized zirconia (PSZ) typically contain lens shaped intra-granular precipitates of the tetragonal phase inside a cubic zirconia matrix, which are fully coherent. [6]

Furthermore, cubic stabilized zirconia (CSZ) [5] and fully stabilized zirconia (FSZ) exist, where only one phase of zirconia is present. This form is used for fuel cells and oxygen sensors. [7]. Also the piezo ceramics lead zirconate titanate (PZT) should be enumerated here, the abbreviation of which is not to be confused with TZP. Its application is mainly related to piezoelectric or ferroelectric effects.

2. Literature

2.1. Fracture mechanics and cyclic loading

This section shall recall briefly the most relevant aspects of linear elastic fracture mechanics for this work. It is used to describe the behavior of cracks in ceramics as well as in metals; however, there is the restriction that the region of plastic deformation close to the crack is small enough. Numerous methods to take into account for larger plastic deformation can be found in the literature, but will not be treated here. A crack in a mechanically loaded material leads to a stress increase at the crack tip, which decreases with increasing distance to the crack tip. The amount of the stress rise depends on the crack length, on the geometrical conditions and on the applied external load. The stress intensity factor K can be used, in order to get a measure which allows to compare cracks in situations where these conditions are different. Common definitions are [5, 8]:

$$K_I = Y\sigma\sqrt{\pi a} = Y_\pi \sigma \sqrt{a} \tag{2.1}$$

Herein σ is the applied stress due to the mechanical load (without crack), a is half the crack length (distance from the center of the crack to the crack tip), Y is a function depending on the crack geometry and the boundary conditions of the mechanical model containing the crack. The index I refers to the opening mode of the crack (mode I: opening normal to the crack plain, mode II: opening by shear in the direction of the crack propagation, mode III: opening by shear transversally to the crack propagation). For surface cracks, a corresponds to the crack length (distance from the crack origin at the surface to the crack tip). Some authors include $\sqrt{\pi}$ of equation 2.1 into the geometry function Y_π. In order to distinguish both definitions, the index π is introduced.

When the stress intensity factor reaches the critical value K_{IC} (the fracture toughness), instable crack propagation occurs, leading to fracture.

When a cyclic load is applied to a part, the stress reaches a minimum σ_{min} and a maximum σ_{max} during one cycle. Thus, it varies about the range ${}_\Delta\sigma = \sigma_{max} - \sigma_{min}$. For a (mode I-) crack contained in the part, the corresponding stress intensity range ${}_\Delta K_I$ can be expressed using equation 2.1:

$$ {}_\Delta K_I = Y(\sigma_{max} - \sigma_{min})\sqrt{\pi a} \tag{2.2} $$

Stable crack propagation in parts subjected to cyclic loading can be described by the Paris law as a function of the applied stress intensity range:

$$ \frac{da}{dN} = C_\Delta K^m \tag{2.3} $$

The left hand side in this equation corresponds to the crack propagation speed with a being the crack length and N being the number of applied load cycles. C and m are empirical constants depending on the material but also on testing conditions such as the applied frequency, mean stress, loading mode or environment. [8] (p. 158)

The applied load amplitude $\sigma_a = \Delta\sigma/2$ as well as the mean stress $\sigma_{mean} = \sigma_{min} + \sigma_a$ (or the applied strain values with the corresponding subscripts) are the main criteria which limit the number of cycles N_f a material can sustain until failure. The applied mean stress is often given indirectly, in terms of the load ratio R:

$$R = \frac{\sigma_{min}}{\sigma_{max}} \tag{2.4}$$

With this parameter, the maximum stress is related to the stress range $\Delta\sigma$ (and to the respective stress amplitude) as follows:

$$\sigma_{max} = \frac{\Delta\sigma}{(1 - R)} \tag{2.5}$$

Equations 2.4 and 2.5 can also be formulated replacing σ by K_I (but keeping the subscripts), when equation 2.1 is introduced.

2.2. Fatigue of metals

A very elaborate overview about the research on fatigue, not only of metals, is given in the books of S. Suresh entitled "Fatigue of materials" [8]. Many recent contributions and reviews about the fatigue of metals were given by H. Mughrabi. For instance, he wrote a detailed review of fatigue in metals related to questions, which are currently subject of research (e.g. very high cycle fatigue, fatigue of ultra fine grained or nanocrystalline metals) [9], or a review of crack initiation caused by fatigue irreversibilities [10].

It is common to distinguish between different fatigue regimes. In the low cycle fatigue (LCF) regime, the ductility of the material controls the lifetime and the material fails after a relatively low number of cycles ($<5 \times 10^4$, [11]).

In this regime, loading amplitudes are high enough for cracks to propagate already during the first fatigue cycles. Plastic deformation is very high, leading to strong hardening or softening and stress-strain hystereses during cycling. Therefore, it is reasonable to express the applied load in terms of strain rather than stress, when focusing on this regime. The plastic strain amplitude and the mean plastic strain are critical in terms of material degradation. When these parameters are controlled, the cyclically induced degradation can be kept constant during the fatigue experiment.

In the high cycle fatigue (HCF) regime, the material strength is more relevant, and a relatively high number of cycles can be reached until failure ($>2 \times 10^6$, [11]). The regime in between the LCF and the HCF range, is described with "finite life" [11]. The overall plastic deformation occurring per cycle in the HCF regime is low and can almost be neglected with respect to the elastic deformation. Thus, a load controlled fatigue test is adequate to maintain constant the critical size for degradation in this regime.

Lifetime diagrams are typically used to describe the fatigue behavior of a material. Herein, the load amplitude (strain amplitude or stress amplitude), which was maintained at a constant value during the experiment, is plotted over the number of sustained load cycles to failure (logarithmic scale) for each sample. When a continuous curve is interpolated using these data points, so called S-N-curves (stress amplitude over number of cycles to failure N) are found. Three examples are schematically shown in figure 2.1.

The continuous curve represents the behavior of materials that harden by strain-aging, such as mild steels. It shows a plateau value for a high number

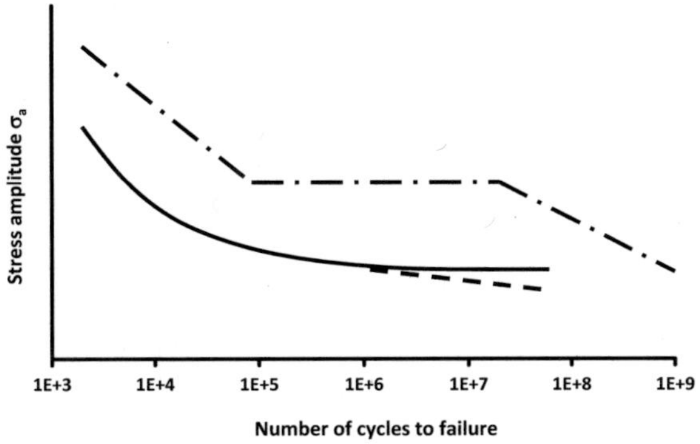

Figure 2.1. Schematic of typical lifetime diagrams in the form of S-N-curves: (Continuous curve) Metallic alloys which strain-age-harden and show an endurance limit [8] (pp.127-128); (Dashed curve) Metallic alloys which do not strain-age-harden and exhibit a continuously decreasing S-N-curve towards high numbers of cycles [8] (pp.127-128); (Dot and dash line) Multi-stage S-N-curve of a high strength steel which decreases after a horizontal segment towards higher lifetimes in the VHCF-regime (beyond 1×10^7 cycles) [12–15].

of cycles. This is called "endurance limit", below which no failure may occur regardless of the number of applied cycles. In contrast, many materials that do not strain-age-harden, e.g. high strength steel or aluminum alloys, exhibit a continuously decreasing S-N-curve for high lifetimes (see dashed curve). [8] (pp.127-128)

A so called multi-stage lifetime diagram is plotted with the dot and dash line. The S-N-curve is horizontal at high lifetimes (from about 1×10^5 to 1×10^7 cycles) and has a negative slope at even higher cycle numbers (beyond about 1×10^7 cycles), see [12–15]. This behavior is typical for high strength steels containing inclusions and can only be revealed, when fatigue experiments are conducted up to the so called very high cycle fatigue (VHCF) regime (more than 1×10^7 cycles also called ultra high cycle fatigue regime, UHCF, [9]). In this regime, (sometimes unexpected) failure can occur, due to very limited local plastic deformation, which is accumulated over millions or billions of cycles [9]. Details will be given in section 2.2.2.

In order to reach the VHCF-regime in a reasonable amount of time, VHCF-fatigue tests are typically conducted at very high frequencies, such as 20 kHz. In this case, cyclic motion is induced by means of a piezoelectric or magnetostrictive transducer, which create longitudinal ultrasonic waves. The resulting mechanical vibrations are amplified using an ultrasonic horn. The geometry of all components including the sample is optimized to obtain a resonant continuum vibration with a maximum stress and strain amplitude in the center of the sample. [16]

Many parameters can have an impact on the lifetime of a component. The most important ones are the loading amplitude, the mean stress (i.e. R-ratio), the environment, the testing frequency, the loading history (constant or variable amplitude, overloads), the shape of the loading curve, the material state (work hardened, annealed, ...), residual stresses, the surface quality (notch effects), the testing temperature, and last but not least the stress state (uniaxial, stress gradients, multiaxial, ...).

2.2.1. Damage evolution - Nucleation and propagation of cracks

When a crack in a component or in a sample grows to a critical length, failure occurs. The most obvious kind of failure is the separation of the fatigued specimen into two parts. Also other criteria appearing at an earlier stage of fatigue can be used to define the end of a component's fatigue lifetime, e.g. an unacceptable loss in stiffness or a decrease in electrical conductivity due to the fatigue induced separation of material.

Small cracks, which grow during fatigue, can be present in the material right after manufacturing. However, mechanisms are described in the literature, which can lead to the formation of cracks, mostly at preferred sights of nucleation. In this context, the motion of dislocations during cycling has an important impact.

The role of dislocations

Slip irreverisiblities of moving dislocations promote slip steps e.g. persistent slip bands (PSB) arising at the surface, which can act as stress concentrators and crack initiation sites. Such irreversibilities are annihilation of dislocations, loss of dislocations at the surface and cross slip. Additionally, in body-centered cubic (bcc) metals, a slip plane asymmetry in tension and compression can lead to small irreversible plastic deformations; i.e., dislocations glide on different glide planes during forward and reverse loading [9, 10].

Depending on the ease of cross slip of dislocations, the cyclic stress-strain (CSS) behavior (evolution of the stress-strain dependence during cycling) depends on the loading history or not. Easy cross slip (so called wavy slip) induces a CCS-behavior, which is independent of the loading history. In contrast, when cross slip is difficult (so called planar slip), the loading history influences the CSS-behavior. As a result, different dislocation distributions are formed during fatigue, namely the following ones for wavy slip: dislocation cells (at high plastic strain), dipoles or bundles of dislocations and persistent slip bands. For planar slip, planar arrangements of dislocations or dislocation groups are formed. In transition regions, mixed structures appear. The formation of these different arrangements depends also on the applied loading amplitude. [9]

Especially in face-centered cubic (fcc) materials, single slip systems (only one slip system is active) are predominately at the surface, and in the bulk multiple slip is required to overcome the constraints by the surrounding grains. [17] (p. 102)

Crack initiation and propagation

Crack initiation occurs usually at the sample surface caused by stress concentrators. The most important ones are surface roughness (notches), surface protrusions due to pronounced slip bands, inclusions, second phases, precipitates, pores, grain and phase boundaries due to the elastic and plastic anisotropy of the microstructure in polycrystals. For HCF loading conditions, up to 90 % of the fatigue life can be determined by crack initiation and the propagation of microstructurally short cracks. [17] (pp. 99-100)

In polycrystals, elastic anisotropies in the microstructure induce local stress concentrations at sites such as twin boundaries, grain boundaries and phase boundaries, which facilitate crack initiation. They are predominant for low and very low strain amplitudes (HCF and VHCF regime). After initiation, the crack propagates following slip bands according to the maximum resolved shear stress. For crack initiation at high strain amplitudes (LCF regime), mainly the plastic slip incompatibility of neighboring grains is relevant (misorientation between the primary slip systems of adjacent grains). [17] (pp. 117-120)

Elastic anisotropy (leading to locally high normal stresses acting on the grain boundary) and plastic incompatibility (leading to dislocation pile-up and dislocation pairing) promote intercrystalline crack initiation. Grain boundaries having a high dislocation density can be considered as slip planes. This is similar to grain boundary sliding at high temperatures, a mechanism which can induce intercrystalline crack initiation without showing pronounced plasticity. [17] (p. 122)

However, according to Suresh [8] (p. 114), grain boundary fatigue crack nucleation is relatively less common in ductile solids without grain boundary particles, creep deformation or environmental influences.

Transgranular crack initiation can appear, when slip systems are activated and dislocation motion is hindered e.g. by grain boundaries, the interface of which is strong enough not to separate. Instead, slip bands can separate leading to transgranular cracking. Preferential sites where this occurs are triple points and intersection points of operated slip bands. [17] (p. 126)

Fracture along crystallographic orientations typically occurs during fatigue of ductile polycrystalline material, where the number of activated slip systems is limited (e.g. difficult cross slip in materials with low stacking fault energy, [18] (p. 25)). In this case, slip lines are sharp straight lines.

During stage I crack growth [mode II crack opening] in ductile solids, single shear appears in the direction of the primary slip system. This leads to a zig-zag crack path, the segments of which are parallel to slip lines. During stage II crack growth [mode I crack opening], simultaneous or alternating flow along two slip systems leads to a planar mode I crack path perpendicular to the far-field tensile axis. [8] (pp. 194-198)

The stress level at the crack tip can be reduced, when the crack faces come in contact with each other, and therefore shield a part of the external load. [As a result, the velocity of crack propagation is reduced.] This phenomenon is called crack closure. The following points promote roughness-induced crack closure [8] (p. 245): 1. low stress intensity factor levels (plastic zone at crack tip smaller than average grain dimension), 2. small crack tip opening (at low $_\Delta K$ and low R) comparable to the average height of fracture surface asperities, 3. coarse grains as well as shearable and coherent precipitates engendering planar crystallographic slip, 4. periodic deflections in the developing crack path induced by grain boundaries and second phase particles, composite reinforcements, abrupt load changes and 5. enhanced slip irreversibility, especially due to slip step oxidation in moist environments.

Crack closure is particularly prominent for near-threshold fatigue, where low amplitudes are applied, at which the crack propagates extremely slowly. [8] (p. 207)

The crack growth rate as a function of the applied stress intensity $_\Delta K$ is represented in figure 2.2. In the Paris regime, between the dashed vertical lines, the crack growth curve (solid line) is described by equation 2.3 and has a constant slope m in the logarithmic plot. For lower stress intensities $_\Delta K$, the crack growth rate decreases rapidly and tends towards a threshold value $_\Delta K_{th}$, below which no crack growth can be detected. At stress intensities beyond the Paris regime, the crack growth rate rises strongly, until the stress intensity reaches a critical value $_\Delta K_c$, at which catastrophic failure occurs (see [8] (pp. 202-203) and [19] (p. E12)). Small cracks can deviate from this behavior and may show particularly high growth rates at low stress intensities (see [8] (p. 292) and [20]). This is indicated with the dotted line and will be treated in section 2.2.1.

Short cracks

Cracks of different size scales show different propagation behavior. Compared to long cracks, which are typically investigated in order to obtain crack growth curves containing the Paris regime (see. equation 2.3), small cracks can show anomalous growth behavior (e.g. higher growth rates, see figure

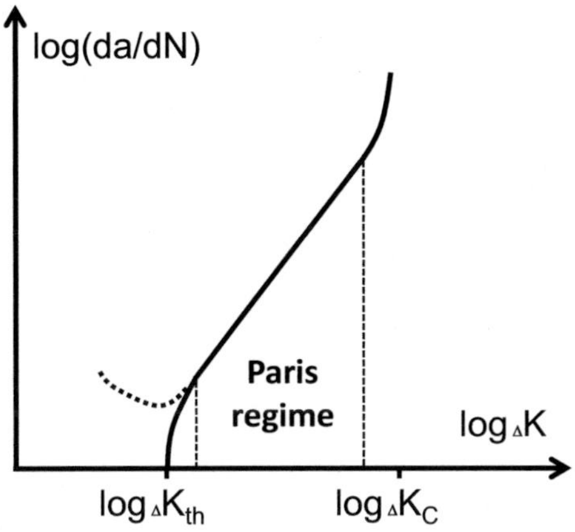

Figure 2.2. Typical crack growth curve for a metallic alloy as a function of the applied stress intensity during fatigue (solid line) derived from [8] (pp. 202-203) and [19] (p. E12). Stable crack growth occurs in the Paris regime. At higher stress intensities, the growth rate increases strongly until critical failure occurs at $_{\Delta}K_c$. At stress intensities below the threshold ΔK_{th}, no crack growth can be detected. Small cracks, however, can show particularly high growth rates at low stress intensities (dotted curve after [8] (p. 292) and [20]).

2.2). According to Suresh, it is common to distinguish between the following kinds of short cracks: 1) microstructurally short cracks 2) mechanically short cracks 3) physically short cracks and 4) chemically short cracks. In these cases, the crack length is comparable to 1) the characteristic size of the microstructure (e.g. grain size), 2) the plastic zone size of the crack, 3) less than one millimeter but significantly larger than the scale of local plasticity, 4) a size scale at which stress corrosion depends on the crack size. [8] (pp. 293-294)

According to Krupp [17], long cracks are larger than 0.5 mm and linear elastic fracture mechanics is applicable to physically short and long cracks. Physically short cracks do not have completely developed plasticity-induced crack closure in contrast to long cracks. Mechanically short cracks propagate mainly in mode I, whereas microstructurally short cracks propagate mainly in mode II and only in parts in mode I.

Taylor and Knott studied the short crack growth behavior of aluminum bronze (CuAl9.5Fe5.0Ni4.5Mn1.25 in weight %) having a grain size in the range of 100 µm [20, 21]. For cracks having a length between 120 µm and 400 µm, they found an elevated propagation rate compared to long cracks. This rate varied according to interactions with the microstructure. It slowed down at a grain boundary, which induced crack branching and reached the long crack growth behavior later on. The resulting crack propagation showed the typical short crack growth behavior with relatively high but irregular growth rates.

Impact of the fatigue frequency

The fatigue testing frequency may have an impact on the lifetime. One reason is that the distance which can be overcome by diffusion depends on the time, and at higher frequencies a larger number of cycles is reached in the same amount of time compared to lower frequencies. Mughrabi points out that frequency effects can be expected, when diffusional processes govern the fatigue behavior. For diffusion controlled fatigue experiments conducted in the HFC (at a low frequency) and VHCF (at a high frequency) regime, a higher lifetime would be expected in terms of cycles to failure for higher testing frequencies, when all the other testing conditions are identical. [22]

Mayer et al. investigated frequency effects experimentally. They found that load controlled tests showed higher local plastic strain when conducted at lower frequencies, encouraging the formation of persistent slip bands. This frequency effect was also dependent on the applied load amplitude and confirmed for descending loads. It was more pronounced starting with a PSB-structure than starting with a dislocation cell structure. [23, 24]

Frequency effects can also be related to corrosion. At lower frequencies, more time per cycle is available for interaction with the environment leading to higher crack growth velocities. Furthermore, rupture of a protective oxide layer due to cycling enhances crack growth. [8]

Stanzl-Tschegg et al. found higher lifetimes in the VHCF regime at 20 kHz than at 100 Hz for tantalum, showing relatively high plastic strain amplitudes. However, for niobium, showing less plastic strain, no difference in lifetime was found for the different frequencies [25]. Also for an aluminum alloy, no difference in lifetime was found for 20 kHz and 100 Hz tests [26]. Furuya conducted experiments on high strength steel at frequencies of 20 kHz and less than 20 Hz. The resulting lifetime diagrams are almost identical [27].

2.2.2. Very High Cycle Fatigue

The traditional lifetime diagram, where the stress amplitude is plotted over the number of cycles to failure (in a logarithmic scale), contains an endurance limit in terms of a stress amplitude, below which no failure occurs (at least for body-centered cubic (bcc) materials). However, it turned out that at cycles far above 1×10^7 cycles (VHCF regime), fatigue failure can arise below the classical endurance limit.

Mughrabi suggests distinguishing between type I materials, which typically are single-phase face-centered cubic (fcc) materials exhibiting surface fatigue cracks and type II materials, such as high strength steels containing non-metallic inclusions, which can show internal fatigue cracking. A further type Mughrabi mentions is a multiphase material such as titanium alloys, which exhibit internal cracking without having inclusions, probably due to elastic-plastic incompatibilities [15]. The VHCF-investigations, which are reported in the literature, typically deal with type I or type II materials.

Type I materials might fail due to surface roughening albeit loaded below the classical endurance limit, more specifically below the PSB-threshold (minimum amplitude to form cyclic strain localization in PSBs). In this case, slightly irreversible random slip, accumulated over a very high number of cycles, is necessary to induce surface roughness, which becomes high enough to act as stress concentrator initiating surface fatigue cracks. [15]

More recent experimental studies show for polycrystalline copper that surface roughening occurs below the PSB threshold (62.6 MPa: minimum loading amplitude at which PSBs are formed within 2×10^6 cycles) and slip lines can arise at stress amplitudes half of the PSB threshold after 1.3×10^{10} cycles (the authors named this value slip band threshold, SB). Samples can be fully covered with slip bands at amplitudes slightly below the PSB threshold after 1×10^{10} cycles. However, failure occurred only at amplitudes of at least 1.5 times the PSB threshold (i.e. 92.2 MPa) after 1×10^{10} cycles. [28]

PSBs (demonstrated using TEM) could even be found by Stanzl-Tschegg et al. after cycling at amplitudes of 25 % below the above mentioned PSB threshold (i.e. 45 MPa) until 2.7×10^8 cycles were reached. The authors named this the VHCF-PSB threshold in order to make the difference to the conventional PSB threshold of 62.6 MPa. They expect that even at lower amplitudes PSBs can be formed when the number of fatigue cycles is increased. For loads below the fatigue limit, small cracks are formed in the VHCF regime, arising from PSBs (as described above), from grain boundaries intersecting with slip bands, or from agglomerated voids or vacancies in the bulk. However, they are not long enough to propagate (30 µm or smaller), and therefore do not lead to failure. Thus, the lifetime diagram is not affected by VHCF damage formed below the fatigue limit, below the traditional PSB threshold or below the VHCF PSB threshold. [29]

Type II materials (high strength steels with inclusions) typically fail due to surface cracking in the conventional fatigue regime (LCF and HCF), whereas a different failure mechanism, namely internal failure originating at a non-metallic inclusion, is observed in the VHCF regime. The smaller the number and the size of inclusions are and the smaller the volume of fatigued material is, the longer lifetimes can be expected in this regime. [15].

VHCF-fracture initiated at inclusions is typically named fish-eye fracture, due to the morphology of the fracture surface. The fish-eye is a more or less circular area appearing on the fracture surface. In the center, this area contains an inclusion surrounded by a small so called optically dark area (ODA). The ODA appears differently on the fracture surface compared to the surrounding material. Several other names can be found in the literature, describing the ODA (coming from optical microscopy), such as fine granular area (FGA) and granular-bright-facet (GBF, coming from scanning electron microscopy).

Some possible explanations for the formation of this kind of fracture are given in a review by Sakai [13]:

1. One model was proposed by Sakai himself. Around the inclusion, fine sub-grains with different orientations are formed during cycling (shown by transmission electron microscopy, TEM). The subsequent formation of micro-debondings and their coalescence leads to the formation of a penny-shaped crack around the inclusion. [13]
2. The following model was proposed by Murakami et al. [30, 31]. Trapped hydrogen around the inclusion induces fatigue crack growth assisted by

hydrogen embrittlement, forming a rough crack surface (ODA) in the vicinity of the inclusion. When this area reaches the size for the material's intrinsic fatigue limit (border of ODA), the crack propagates without assistance of hydrogen, until the border of the fish-eye area (around the ODA) is reached. This border is associated with the start of unstable static fracture at the end of the fatigue life. Calculations based on fracture mechanics show that most of the fatigue life is spent in crack initiation during formation of the ODA. [30, 31]

3. A further model was proposed by Shiozava et al. [14]. Around the inclusion, microcracks are induced by decohesion of dispersed spherical carbide from the matrix. They join each other and form the rough portion of the fracture surface around the inclusion. After this period of possibly very slow crack growth, the crack propagates independently of the microstructure as an ordinary crack and the fish-eye is formed. [14]

According to all models cited in section 2.2.2, the lifetime in the VHCF regime is governed by crack nucleation. However, McEvily et al. [32] believe that subsurface cracks, which are typically the origin of VHCF-failure for the alloy Ti-6Al-4V, are initiated early in life. The authors suggest that such subsurface cracks are initially arrested because of crack closure. Due to wear (in absence of oxygen), the level of closure can be reduced during cycling, until the crack starts to propagate at a very slow rate. Thus, further closure is developed. The optically dark area, which is typically formed around a failure inducing subsurface inclusion, is a rough region. McEvily et al. attribute its formation to the degradation of roughness induced closure by wear, and assume that this process takes the largest portion of the very high cycle fatigue life. [32]

2.2.3. Scaling and size effects

Scaling and size effects must be considered, when the dimensions of an investigated structure are reduced. Nowadays, it is not common to distinguish these terms in the literature. Mainly the term "size effects" is used to describe both phenomenons. However, it is suggested to assign these terms as follows, in order to categorize changes of dominating mechanisms caused by reduced sizes.

When the reduction of dimensions has a remarkable influence on some phenomenon, but the mechanism and the law describing this phenomenon do not change, we call this a scaling effect. For instance, the surface to volume ratio of an object increases, when it is scaled down in size (while keeping the ratio of all dimensions constant). Thus, the thermal exchange of this object with its environment is enhanced, whereas the law and the mechanism describing the heat transfer stay the same. Scaling effects describe continuous changes, which occur when the size is scaled down continuously.

However, at some specific size, a sudden (discontinuous) change in behavior might be observed, when the dimensions are scaled down. At this point, a law or a scaling law, which might be used to describe the regime of continuous change in behavior at a larger size scale, is not sufficient any more to describe the material behavior.

Two main reasons can cause such a transition point in material behavior at a specific size: first, an interference of the object (or sample) size with a size related to discrete elements determining the material behavior (e.g. the sample size reaches the order of the distance between flaws), and second a change of mechanism determining the material behavior (e.g. reduced plastic deformation in thin films compared to bulk material, because of constrained dislocation motions). It is suggested to name such effects size effects, which appear when at least one dimension of an object reaches a specific size.

Concerning fatigue at the small scale, size effects were investigated by Schwaiger and Kraft cycling Ag films of different thicknesses (0.2 µm to 1.5 µm). For the thinner films, the authors found a steep increase of the stress amplitude needed to induce damage after 3.8×10^6 cycles. They concluded that this indicates a fatigue behavior which is affected by dimensional constraints and leads to an enhanced fatigue resistance for thinner films. [33]

Furuya investigated high strength steel up to the gigacycle regime [27] at different size scales. He found different lifetimes for different sample geometries. The latter were designed in a way to obtain different sizes of volume subjected to the highest stress. Samples having the smaller volumes showed a higher resistance to fatigue and smaller crack inducing inclusions, whereas samples with larger volumes failed earlier caused by larger inclusions. Furuya reasoned that this phenomenon is due to an increase in probability for having a large inclusion, when the investigated material volume is enlarged.

One could categorize the observations of Furuya in terms of size and scaling effects as follows. At the large scale, the material shows the typical fatigue behavior leading to a lifetime diagram with a curve relating the applied stress amplitude to a certain number of cycles until failure. When the sample size is scaled down to a sufficiently small dimension, a size effect can be observed. A certain stress amplitude cannot be related to a specific number of cycles to failure anymore, because a high scatter in lifetime (more than one decade) occurs. The sample is small enough, not to contain an inclusion of the most critical size of the material at an elevated probability.

When the sample size (volume) is still scaled down more and more, the fatigue strength increases accordingly, because of the reduced probability of having large inclusions in the sample. This continuous increase of fatigue strength and decrease of inclusion sizes inside the sample is a scaling effect.

Besides the presented size effects, which are always related to the reduction of a sample dimension, changes of mechanisms can sometimes be purely related to the size of the microstructure (e.g. different mechanisms of plastic deformations for small grain sizes compared to coarse grains). In this case, when the sample dimension is not a reason for the change of material behavior, it is suggested not to assign the phenomenon to scaling or size effects, but to categorize it as a purely microstructural effect.

Some examples for such microstructural effects are given in the following, which are neither size nor scaling effects (only the size of the microstructure scales down, not the size of the sample) according to the proposed definition. They can play an important role for micro samples, since such specimens typically have a fine microstructure.

An example is the change of mechanisms enabling plastic deformation. In coarse grain metals, dislocations can nucleate, glide, interact and pile-up inside a grain and the resulting stress fields influence the dislocation activity in neighboring grains. When the grain size is scaled down towards the nanocrystalline size, dislocation behavior is changed. Thus, plastic deformation can be generated by other mechanisms, such as grain coarsening, grain boundary sliding, grain rotation, or twinning (see e.g. [34]). The plastic deformation of very clean single crystal whiskers, having dimensions in the nm regime, can be nearly restricted, pushing the strength of material towards the theoretical strength [35].

2.3. Fatigue of ceramics

When ceramics are subjected to cyclic loads, i.e. fatigue experimentation, an observed time dependent degradation is not necessarily caused by the variation of the load over time. E.g., for glass it was shown that the time dependent degradation mechanism inducing failure during cycling is the same which occurs during application of a static load (smaller than the strength). This mechanism is related to stress corrosion and is called subcritical crack growth. Besides glass, typical ceramics exhibiting subcritical crack growth are alumina, mica, silicon nitride or zirconia.

According to [36], a degradation effect of ceramics, which does only occur during cycling and not during pure application of a static load, was first reported in 1986 for compression (polycrystalline alumina, [37]) and in 1987 for tension (Mg-PSZ, [38]).

Due to the brittle behavior of ceramics, the load leading to failure depends strongly on the size and on the distribution of flaws contained in the tested material. Thus, the load leading to failure of ceramics typically shows a high scatter. This is valid for spontaneous failure as well as for time dependent failure.

2.3.1. Crack propagation mechanisms

Equation 2.3, showing the Paris law, was initially introduced to describe the behavior of metals. Even though the governing mechanisms are different, time dependent crack propagation in ceramics, subjected to a static or to a dynamic load, can be expressed using the same formulation. It might be therefore, that the time dependent degradation of ceramics under a static load is sometimes called "static fatigue" in the literature. However, in the present work it is preferred to use the term fatigue only in the context of cyclic loads. In order to clearly distinguish between metals and ceramics, a different notation is used to express the Paris law for ceramics. Furthermore, a common normalization of ${}_{\Delta}K_I$ by K_{IC} is introduced to calculate the crack growth rate v, leading to:

$$v({}_{\Delta}K_I) = \frac{da}{dN} = A_{\Delta}{K_I}^n = A^* \left(\frac{{}_{\Delta}K_I}{K_{IC}}\right)^n \tag{2.6}$$

Note that (in contrast to A^*) the crack growth exponent n is not affected by the normalization. The above equation is formulated for cyclic loads, but it also holds in case of time dependent failure due to static loads, when N is replaced by the time to failure t and ${}_{\Delta}K_I$ is replaced by K_I.

Experimental observations showed that varying the R-ratio (i.e. mean-stress or maximum stress) at identical load amplitude usually has a much stronger impact on crack growth of ceramics than vice versa. In equation 2.6, the effect induced by the R-ratio is included in the parameters A and A^*. In order to separate the impact of K_{max} (related to σ_{max}) and of ${}_{\Delta}K$, Liu et. al. suggested for 3Y-TZP to split the exponential law of equation 2.6 into two parts, one depending on K_{max}, the other depending on ${}_{\Delta}K$ [39, 40]. This can be formulated using equations 2.6 with 2.4 and splitting the crack growth exponent into two components k and l with $n = k + l$ (see also [5] (pp. 115-116)):

$$\frac{da}{dN} = A_{\Delta}{K_I}^{(k+l)} = A(K_{max}(1-R))^k{}_{\Delta}{K_I}^l = A(1-R)^{(n-l)}{K_{max}}^{(n-l)}{}_{\Delta}{K_I}^l \tag{2.7}$$

Liu et. al. found the exponent k (related to the maximum stress intensity) to be one order of magnitude higher than l (with $l = 2$).

Subcritical crack growth usually is associated with chemical reactions at the crack tip (e.g. induced by water molecules). Different stages of subcritical crack growth (observed in crack propagation experiments) are visualized in figure 2.3. They are explained in the literature by thermally activated stress corrosion models of ceramics (e.g. in the presence of water). For instance, Chevalier et. al. [36] fitted models developed by Lawn to their experimental results on 3Y-TZP, and showed a good agreement for the following mechanisms:

Stage I, which can be described by equation 2.6 (modified for static loads): Mainly governed by chemical reaction. It was fitted in [36] to a corrosion law describing the chemical reaction between zirconia and the water molecules. Therefore, Chevalier et al. used the bond spacing of zirconia (derived from the lattice parameter), the mechanical energy related to K_I and the activation energy for the reaction with water (derived by double torsion experiments at different temperatures) as well as the activation area and the fracture surface energy as two fitting parameters.

Stage II: The transport of corrosive molecules to the crack tip, prior to the chemical reaction mentioned above, limits the crack propagation velocity and leads to a constant rate. This stage was fitted to a diffusion law which depends mainly on the water pressure, the mass of the water molecules and the crack opening displacement (the latter was used as the fitting parameter).

Stage III: Fracture like in vacuum conditions. The experimental data was fitted by [36] using the law from stage I adapted for vacuum conditions (activation energy for bond fracture without chemical reaction).

For some ceramics a threshold K_{Ith} exists, below which no crack growth occurs. In the work of Chevalier et al. [36], K_{Ith} was lower in water than in air (see curves a and b in figure 2.3 (Right)). This supports the mechanism of stress corrosion like the following observations do, which are shown in figure 2.3 (Right): Experiments in water, silicon oil and vacuum do not show stage II in contrast to experiments in air. The crack growth curve section of stage III from experiments in air lies on the curve found under inert conditions (silicon oil and vacuum, curve c). The crack growth curve of experiments in water is shifted towards lower K_I-values with respect to experiments in air, showing however the same slope in stage I. The same is valid comparing water at ambient temperature (curve a) to water at 75 °C (curve b). The phenomenon of subcritical crack growth is affected by the crack growth resistance K_{IC}. Therefore, the so called R-curve behavior must be taken into account.

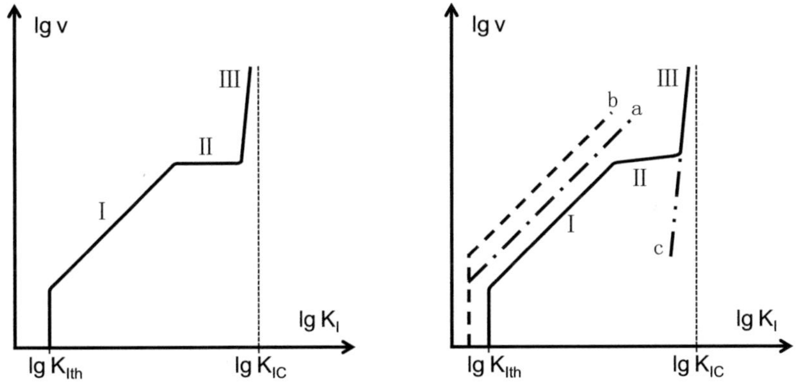

Figure 2.3. (Left) Crack growth curve for static loads showing three stages of subcritical crack growth (see e.g. [5] (p. 78)); (Right) Effect of temperature and media on subcritical crack growth shown in a sketch according to results presented in [36]: solid line - ambient temperature in air, curve a - water at 25 °C, curve b - water at 75 °C, curve c - oil and vacuum.

Ceramics without R-curve behavior have a constant crack growth resistance (ideally brittle behavior: the crack growth resistance K_{IC}, at which instable fracture occurs, is independent of the crack length). In contrast, the crack growth resistance of ceramics with R-curve behavior increases with higher crack extension $\Delta a = a - a_0$. The stress intensity factor K_σ as a function of the crack length a according to equation 2.1 is plotted in figure 2.4 (Left) for different applied stresses σ (with $\sigma_a < \sigma_b < \sigma_c$). In addition, two different crack growth resistance curves K_{R1} and K_{R2} are plotted. They describe the levels of stress intensity, which are necessary to prolongate a crack having the initial length a_{01} and a_{02} respectively towards a new length a (both cracks have the same geometry function Y from equation 2.1). K_{th} is the threshold value, below which both cracks do not grow (K_{th} could also be different for both curves, e.g. if they would be assigned to different materials).

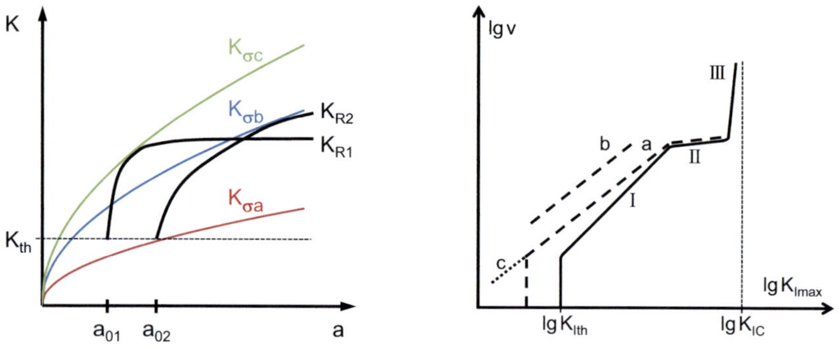

Figure 2.4. (Left) Stress intensities K_σ over the crack length a for three different applied loads σ with two types of R-curve K_R; (Right) Effect of cycling on crack growth (ambient temperature in air): solid line - constant load corrected for comparison with cycling, curve a - cyclic load, curve b - cycling at lower frequency [36] or lower R-ratio [40] than for curve a, curve c - natural flaws (in contrast to macrocracks for curve a) [41].

The applied stress level σ_a leads to stress intensities (red line designated $K_{\sigma a}$) which lie below both crack resistance curves. Thus, the cracks a_{01} and a_{02} do not grow. When a higher stress σ_b is applied, a_{01} grows until it reaches a length a at which the stress intensity $K_{\sigma b}$ is equal to the crack resistance K_{R1} (intersection between both curves). This is a stable point at which the crack stops growing. In contrast, the curve $K_{\sigma b}$ meets curve K_{R2} at the tangent point, which is an unstable point and thus related to catastrophic failure (starting from this point, crack propagation leads to stress intensities, which are higher than the crack resistance). $K_{\sigma a}$ is the limit at which unstable crack propagation occurs also for K_{R1}. In summary, instable crack growth starts, when the applied stress intensity K_I is equal to K_{IR} and the following condition is fulfilled: [5] (p. 23)

$$\frac{\partial K_I}{\partial a} = \frac{dK_{IR}}{da} \tag{2.8}$$

The two examples K_{R1} and K_{R2} were chosen to illustrate different types of R-curves. K_{R1} is steep at the beginning whereas K_{R2} increases more slowly over a larger range. Several parameters having an impact on the R-curve characteristics are described in [5] (pp. 53 ff.): the initial crack length a_0, the flaw type (natural flaws or macrocracks), the specimen type (3-point bending, 4-point bending, uniaxial tension), and the material. For some materials, different grain sizes lead to different R-curves. Thus, the relation between the stress intensity K and the crack propagation $\Delta a = a - a_0$ is not a unique material property.

Different reasons causing the R-curve behavior (increase in crack growth resistance by a shielding term K_{Ish}) are given in the literature, which are mainly:

- The interaction between crack surfaces behind the crack tip, so called bridging effects, can work against crack propagation. These are unbroken microstructural sections (bridges) connecting both crack surfaces, frictional forces between rough crack surfaces in contact, or wedged segments between both crack surfaces.

- Microcracks and crack branches forming in front of the crack increase the energy which is needed for crack propagation.
- Phase transformation (e.g. transformation from the tetragonal to the monoclinic phase in ZrO_2) induced by the stress peak in front of the crack tip leads to a volume expansion. Thus, a compressive residual stress develops in the vicinity of the crack tip and works against crack propagation.

As a result, the stress intensity factor at the crack tip K_{Itip}, which governs crack propagation, can be described as follows:

$$K_{Itip} = K_I - K_{Ish} \tag{2.9}$$

Herein K_I is the stress intensity factor related to the applied stress σ according to equation 2.1 and K_{Ish} is the shielding term related to the described mechanisms. The shielded portion of K_I namely K_{Ish} cannot contribute to crack prolongation and is therefore equivalent to an increase in crack growth resistance.

In comparison to subcritical crack growth at constant loads, for some ceramics (not all ceramics show different behavior under static and cyclic loads) cyclic loading leads to lower crack growth exponents n (lower slope) and to higher crack growth rates (shift towards higher crack growth velocities) in a plot such as figure 2.3 (Left) (plot over K_{Imax} for the cyclic condition). This observation is mostly explained by a reduction of toughening mechanisms due to cycling.

The direct comparison between static and cyclic conditions is not straight forward, since K_I changes periodically in the cyclic case. Therefore, Chevalier et. al. [36] calculated a reduced crack propagation rate out of their experimental results for static load conditions, by integrating over a sinusoidally varying stress intensity as it was applied in the cyclic experiments. The result is shown in a qualitative way in figure 2.4 (Right). The solid line represents the reduced crack growth rate, found by [36] applying the described method (which corrects for comparison with cycling). In stage II this curve lies on the dashed line, suggesting that the condition of the cyclic experiment corresponds to a monotonic loading in this regime. According to Chevalier et al. this can be explained by a crack advance per cycle which is remarkably larger than the disturbed zone around the crack tip. In stage I the slope under cyclic conditions (curve a) is lower than under static conditions, and the crack growth rates are higher due to cyclic degradation.

Further effects described in the literature with respect to the cycling are drafted in figure 2.4 (Right): lower frequencies or lower R-ratios at the same maximum stress lead to a shift towards higher crack growth rates (curve b). Natural flaws show the same crack growth exponent as artificially introduced macrocracks do. However, they propagate at stress intensities far below the threshold of macrocracks, i.e., their threshold is lower or does not exist.

Several explanations for the degradation of toughening mechanisms during cycling exist. For instance, bridges between the crack surfaces can degrade during cycling by frictional processes [42], [41], [5] (p. 118). A model from Jacobs et. al. suggests an equilibrium between the degradation of bridges and the formation of new bridges due to crack propagation [36, 43].

In ceramics with phase transformation caused by mechanical loads, a larger transformed zone around the crack tip was observed for purely subcritical crack growth compared to cyclic loads (same maximum load for both tests) [5] (p. 122). Hoffman et al. explain such observations in PSZ-ceramics with a previous frictional degradation of bridges between matrix and precipitates, leading to a reduced strength of these bridges. Thus K_{Itip} is also reduced, which causes a smaller transformation zone size [41].

Liu et. al. [44] assume microcracking, which starts at preexisting flaws, to be the main fatigue mechanism in 3Y-TZP. Furthermore, they suggest that a crack nucleation in the bulk would be favored by internal stresses due to thermal mismatch appearing during manufacturing, elastic anisotropy or phase transformation, which could enhance fatigue damage. The authors claim that plasticity, related to phase transformation, is less important for 3Y-TZP and only relevant for very low stress ranges in this material. However, in Mg-PSZ having a larger capacity of transformation, phase transformation and microcracking are active at all stress ranges with a shift towards transformation mechanisms at high stress ranges. Liu et. al. explain that internal stresses formed by the phase transformation (during stress controlled cycling) oppose the forward transformation (tetragonal to monoclinic) and assist the reverse transformation (monoclinic to tetragonal). At the same time, cyclic creep occurs, because of a smaller transformation stress in tensile direction compared to the compressive direction.

The impact of phase transformation on bending experiments was analyzed in [45] with the focus on Mg-PSZ. Due to spontaneous and time dependent phase transformation, the stress distribution inside the bending bar is changed (reduced stress in the outer fiber). As a result, the Weibull-diagram is expected to have a positive curvature instead of a straight line (for quasistatic tests such as three point bending or tensile tests as well as for lifetime tests at constant loads). According to the authors, the reduction of the outer fiber stress (leading to elevated loads on inner fibers) is more pronounced for the lifetime tests.

Time dependent phase transformation could be related to sequentially transforming particles starting with the most preferentially oriented ones [46], which induce microcracking and internal stresses. Finlayson et. al. could describe the time dependent tensile creep strain by a power law (product of the tensile stress to the power of a material parameter and the time to the power of a second material parameter - the material parameters depend on the microstructure and on the phase contents in the material).

Liu et. al. [44] found the cyclic deformation in Mg-PSZ and in 3Y-TZP to be very stress rate and frequency dependent. At low stress rates phase transformation induced plasticity is favored, whereas it can be suppressed at very high rates leaving only microcracking as the active mechanism (which is usually rate independent).

Schmitt et. al. found for PSZ-material that R-curve effects and related shielding may be negligible for specimens with natural flaws (i.e. without introduced macrocrack), because stage I of the crack growth curve extended far below the threshold for macrocracks towards much smaller crack growth speeds without showing a threshold [47]. Furthermore, they point out that for Mg-PSZ the R-curve increases more gently for cyclic loads and reaches much lower K_I-values than the R-curve for static loads [47], [48].

2.3.2. Statistic evaluation of fatigue data

In the present work, a statistical procedure was used to generate crack growth curves combining quasistatic three point bending strength data with data from cyclic bending tests, according to a method applied by Fett, Munz et al. [49]. The latter is explained in this section, including some fundamental aspects of the statistics behind, which can be found for instance in the book of Munz and Fett [5].

According to the weakest-link approach the most critical flaw (in terms of size and solicitation) in the component induces failure. Flaws are considered to be cracks. Consequently, the flaw size a and the strength σ_c are related by the fracture toughness K_{IC} according to equation 2.1. Thus, the distribution of the strength is related to the flaw distribution. Assuming a flaw density z for the material, the average number of flaws in a component is proportional to its volume V. The probability for a certain number of flaws in one component follows the Poisson distribution with the parameter zV. The flaw sizes a can be described by a distribution with the distribution function $F(a)$. One can calculate the probability $P(a < a_c)$ that a component contains only flaws smaller than a critical size a_c, using $F(a)$ and the Poisson distribution as follows:

$$P\left(a < a_c\right) = \exp\left(-zV\left(1 - F\left(a_c\right)\right)\right) \tag{2.10}$$

Since flaw size and strength are related, a component will fail at stresses $\sigma \geq \sigma_c$ with the complementary probability of equation 2.10. Thus, the distribution function for strength values $F\left(\sigma_c\right)$ is obtained, subtracting the right hand side of equation 2.10 from 1 and replacing a_c by a. Introducing an appropriate crack size distribution function $F(a) = 1 - (a_0/a)^{(r-1)}$, where r is a material specific parameter and a_0 normalizes for $F(a \to \infty) = 1$, the strength distribution can be expressed by the following Weibull-distribution:

$$F\left(\sigma_c\right) = 1 - \exp\left(-\left(\frac{\sigma_c}{\sigma_0}\right)^m\right) \tag{2.11}$$

In this relation, σ_0 and m are the Weibull-parameters. They can be determined graphically applying two times the logarithm to equation 2.11 and plotting $\ln \ln(1/(1-F))$ over $\ln \sigma_c$ for a set of strength data. For this purpose, F_i is determined for each value $\sigma_{c,i}$ by $F_i = (i-0.5)/n$ or by $F_i = i/(n+1)$. Herein $i = 1$ for the smallest, $i = 2$ for the second smallest,..., $i = n$ for the largest σ_c. The parameter m is the slope of the resulting line and describes the scatter of the data (higher m for lower scatter). Since m depends slightly on the number of measurements, it is common to calculate a corrected parameter m_{corr}, which does not depend on the amount of data points. A more precise way to determine the parameters m and σ_0 for a given data set is the maximum-likelihood method. In this case, the parameters are adjusted to the values, for which the probability to obtain the given data set is the highest. Whereas m depends only on a parameter related to the flaw size distribution ($m = 2(r-1)$), σ_0 contains the parameter a_0 of the flaw size distribution and the volume of the tested component:

$$\sigma_0 = \frac{K_{IC}}{(zV)^{(1/m)} Y_\pi \sqrt{a_0}} \tag{2.12}$$

Thus, the larger the volume of a component loaded with a stress σ is, the higher the probability for failure is. For two components having the same failure probability but different volumes V_1 and V_2, the ratio of their strengths is given by:

$$\frac{\sigma_{c,1}}{\sigma_{c,2}} = \left(\frac{V_2}{V_1}\right)^{(1/m)} \tag{2.13}$$

This relation can be used to transfer results from specimens to components. For this purpose V_1 and V_2 are replaced by the effective volume of the component and the specimen respectively. The effective volume takes into account the inhomogeneous stress distributions. The volume dependence and the exponential law in equation 2.11 are due to the Poisson distribution which is used to determine the probability for a certain amount of flaws in a component having the volume V. When surface cracks are responsible for failure, the above considerations are the same and the volumes V must be replaced by the respective surfaces S.

In the following, it is assumed that the same flaws (having a size a_i) which determine the strength σ_c, induce failure related to time dependent degradation, when the applied load is smaller than σ_c. The degradation can be caused by a constant load σ (static load) or by an alternating load with an amplitude $_\Delta\sigma$ (dynamic load). The critical flaw having the initial size a_i grows over time (following the relevant crack growth law) to a size a_c, where failure occurs. Since the (cyclic or static) load is constant during a test, K_I grows in the same way according to equation 2.1 (in which a possible change of Y_π during crack growth is neglected). Thus, the crack growth rate v can be obtained in function of K_I using the derivative $(da/dK_I)_{\sigma=const.}$ of equation 2.1. Introducing the initial stress intensity K_{Ii} which corresponds to a_i and is present at the moment of load application, one obtains:

$$v(K_{Ii}) = -\frac{2K_{Ii}}{Y_\pi^2\sigma^2}\frac{dK_{Ii}}{dt} \tag{2.14}$$

Herein, t is the time to failure (time of growth from a_i to a_c). For cyclic loads, the time t is replaced by the corresponding number of cycles to failure N_f. Furthermore, $_\Delta K_{Ii}$ and $_\Delta\sigma$ are used instead of K_{Ii} and σ. Applying logarithmic derivations as well as the relation $_\Delta K_{Ii}/_\Delta\sigma = K_{IC}/\sigma_c$ (the corresponding equation for static loads is $K_{Ii}/\sigma = K_{IC}/\sigma_c$), one obtains the cyclic crack growth rate:

$$\left(\frac{da}{dN}\right)_{ik} = -\frac{2K_{IC}^2}{N_{fk}\sigma_{ck}^2 Y_\pi^2}\frac{d\log\left(_\Delta\sigma/\sigma_c\right)}{d\log\left(_\Delta\sigma^2 N_f\right)} \tag{2.15}$$

The index k can be ignored considering a continuous function. It was introduced to explain below the numerical treatment of the data acquired in this work. (An equivalent relation exists for static loads.) According to equation 2.15, the crack growth rate at the moment of load application (start of time dependent degradation) can be determined, when the lifetime data ($_{\Delta}\sigma$ and N_f), the strength data σ_c as well as K_{IC} and Y_π are known. K_{IC} can be obtained from fracture toughness tests and Y_π is estimated to be 1.3 (for semicircular surface flaws). The other values are obtained conducting N_L lifetime tests and N_m fracture tests on samples coming from the same batch (containing flaws from the same population). The respective distribution functions F_L (lifetime values) and F_S (strength values) can be approximated calculating

$$F_{Lk} = \frac{k}{N_L + 1} \qquad F_{Sl} = \frac{l}{N_S + 1} \tag{2.16}$$

The index k numbers the obtained lifetime values N_{fk} ($k = 1, 2, ..., k = N_L$) where $N_{f1} < N_{f2} < ... < N_{fN_L}$. Accordingly, l numbers the strength values σ_{cl}. Now, each lifetime value N_{fk} can be assigned to the strength value σ_{ck}, for which the following condition is fulfilled:

$$F\left(N_{fk}\right) = F\left(\sigma_{ck}\right) \tag{2.17}$$

When a different number of lifetime values and strength values were gained, σ_{ck} is determined by linear interpolation between the two strength values having the probabilities F_{SL}, which are the closest to F_{LK}. In the present work, for all batches more strength data than lifetime data was available. An alternative interpolation method is proposed by [50] (p. 67). The strengths σ_{ck} can be calculated using the Weibull-distribution of the strength values.

To do so, equation 2.11 is solved for σ_c and $F(N_{fk})$ is introduced applying equation 2.17. All determined pairs of strength and lifetime are used to draw $(\log(\Delta\sigma/\sigma_c))$ versus $(\log(\Delta\sigma^2 N_f))$. An appropriate interpolation gives the second fraction from the right hand side in equation 2.15 (in our case this fraction corresponds to the slope of a linear fit). Since all data points are used to obtain this term, the contained parameters do not have the index k in equation 2.15. However, the first fraction on the right hand side of the equation contains only one pair of strength and lifetime (indexed with k), which is found by the described method. Thus, the initial crack growth rate can be calculated for each lifetime test k as a function of ΔK_{Iik}, using equation 2.15 and the relation $\Delta K_{Iik} = (\Delta\sigma K_{IC})/\sigma_{ck}$. Fitting through all data points, the parameters from equation 2.6 are derived, in particular the crack growth exponent n. Data from specimens with spontaneous rupture and from specimens without rupture are included to calculate F_{Lk}, but they are not used to calculate crack propagation rates.

2.4. Aluminum Bronze

Micro molded aluminum bronze is the metallic alloy which was investigated in this work. Some introductory remarks on this material were already given in chapter 1. In section 2.4.1, the microstructure of aluminum bronze (on the macroscopic scale) will be presented, which has an important impact on the material behavior. The main focus is set on alloys having a chemical composition of CuAl10Ni5Fe4 (numbers given in weight %) or close to it, which corresponds to the nominal composition of the investigated micro molded aluminum bronze.

A short description of the micro molding process, which was applied to produce the micro samples out of aluminum bronze, is given in the subsequent section. Studies on their microstructure and on the fatigue behavior will be summarized briefly.

2.4.1. Microstructure and properties

Analyses of the chemical composition, the morphology and the crystallography of the phases contained in CuAl10Ni5Fe5 aluminum bronze were done for example by Hasan et al., Culpan and Rose as well as Feest and Cook. Their results are summarized in the detailed book of Meigh [2] (pp. 300 ff). The identification of the different phases is not always the same from author to author and not all the authors identified the same number of phases in the alloy. The most extensive and systematic work seems to be the one of Hasan et al. and therefore serves as a reference in the following.

A tabular description of the phases is given in table 2.1 according to Hasan et al. [51]. The β phase is only mentioned in its martensitic form (sometimes called retained β or β'), which is the relevant appearance at room temperature. It appears, when the high temperature β phase (solid solution bcc containing Ni, Fe, and sometimes Mn in solution) is not completely transformed into other phases, such as the eutectoid $\alpha + \kappa_{\text{III}}$. It is susceptible to corrosion, leads to higher tensile strength and lower elongation. The cooling rate and the content of Al and Ni govern mainly, whether the β phase is retained at room temperature or not. [2] (pp. 306-312).

According to Weill-Couly et al., complex aluminum bronze (containing Al, Cu, Ni, and eventually Fe) is resistant to corrosion in terms of dealuminifiaction in sea water, when the β-phase is decomposed at least partially into $\alpha + \kappa_{\text{III}}$ [52]. However, massive β without surrounding protective α and κ_{III} is susceptible to corrosion. The reason for the de-aluminification of β in duplex aluminum bronze (CuAl) is that β is more anodic than α.

	Cu/Al/Ni/Fe [wt%]	morphology	lattice [Å]	cryst. structure
α	85.8/7.2/ 3.0/2.8	-	3.64	Cu-rich fcc matrix
β	(20.2/28.1/ 35.1/14.0)	can contain prec.	2.85	martensite (3R or 2H) prec. NiAl based (B2)
κ_{I}	10.5/9.3/ 3.5/72.2	dendritic-shaped 20..50 µm in center of α grain contains Cu-rich prec.	-	no single cryst., contains: Fe-rich solid solution (bcc) Fe3Al (DO3) FeAl (B2)
κ_{II}	12.1/12.3/ 8.0/61.3	dendritic-shaped 5..10 µm	5.71	Fe_3Al based Ni, Cu, Mn, subs. for Fe Si subs. for Al DO3 Schnoeflies
κ_{III}	17.0/26.7/ 41.3/12.8	lamellar/globular degenerate lamel.	2.88	NiAl based (B2) Fe, Cu, Mn subs. for Ni
κ_{IV}	2.6/10.5/ 7.3/73.4	globular sizes $< 2\,\mu m$	5.77	Fe_3Al based, similar to κ_{II} some internally twinned

Table 2.1. Phases in a nominal CuAl10Ni5Fe5 aluminum bronze according to Hasan et al. [51]. The specific compositions were determined with X-ray diffraction techniques using a SEM with EDX detector and a TEM with EDAX detector. Contents of Si and Mn are omitted. The contents given for the martensitic β phase concern the precipitates, which can arise and are therefore given in brackets. Ordered crystallographic structures, such as ordered bcc B2 and DO3, are given in brackets. Abbreviations: prec. precipitates, lamel. lamellar, subs. substituting.

It is probable that internal stresses aggravate this phenomenon. [2] (pp. 239-240), [52]

The κ_{I} phase was found by Hasan et al. in an alloy with an Fe-content of 5.1 % but not in an alloy with a lower Fe-content of 4.4 % (values in weight %). [2]

For Al-contents higher than 11 %, an additional γ-phase can appear at temperatures slightly below 600 °C. It forms a eutectoid with $\alpha + \kappa_{\text{III}}$ and is susceptible for corrosion albeit resistive to wear. [2] (pp. 300, 307)

The melt starts to solidify at about 1080 °C and the temperatures at which phases appear depend strongly on cooling rates and on the elemental composition of the alloy. (A variation of Fe from 4.38 % to 5.09 %, Ni from 4.35 % to 5.1 % and Al from 8.9 % to 9.37 % leads to a change of up to 110 °C of the temperatures at which phases develop). The equilibrium state can only be achieved by quenching at high temperature with subsequent annealing at a selected temperature, but not by slow cooling. [2] (pp. 294-296)

Mechanical properties depend on the composition, thermal treatment and manufacturing process of the alloy. Meigh therefore distinguishes between cast and wrought aluminum bronze. As an example the properties of the cast CuAl10Ni5Fe5 alloy are given in the following, the composition of which is close to the alloy investigated in the present work. Young's modulus: 116 GPa to 124 GPa; Shear modulus: 45 GPa to 48 GPa; Poisson's ratio: 0.3; Tensile strength: 551 MPa to 690 MPa; 0.2 % strength: 250 MPa to 280 MPa; Elongation: 7 % to 13 %; Hardness Brinell: 140 to 150; Fatigue limit at 1×10^8 cycles: 190 MPa to 220 MPa; Coefficient of thermal expansion from 0 °C to 100 °C: $16.3 \times 10^{-6}\,\mathrm{K}^{-1}$; Specific heat capacity: $434\,\mathrm{J\,kg^{-1}\,K^{-1}}$; Thermal conductivity at 20 °C: $38\,\mathrm{W\,m^{-1}\,K^{-1}}$ to $42\,\mathrm{W\,m^{-1}\,K^{-1}}$. [2] (pp. 16-17, 20, 25, 31)

2.4.2. Micro investment casting

Baumeister et al. [53] used aluminum bronze as an alloy for the development of micro investment casting, which is a lost-wax lost-mold process allowing for manufacturing of details in the sub-millimeter range. It uses plastic models of the samples mounted on a feeding system out of wax. The assembly is embedded in a ceramic slurry and sintered. Thus, the plastic is pyrolized leaving the negative of the part in the mold [53]. In contrast to macro parts, the filling of the mold is enhanced by preheating the mold and by applying a force on the melt. Therefore, the mold is evacuated, before it is filled with the melt. Subsequently, air pressurized at up to 4 bar acts on the melt (vacuum pressure casting), or a pressure of 20 bar to 25 bar is induced by centrifugal forces (centrifugal casting) [54]. When the mold is cooled down to room temperature, it is then destroyed in order to access the samples, which are cleaned and separated. [53]

The overall microstructure of the samples was investigated in detail by Rögner, Baumeister and Okolo. It can vary between batches from very fine to very coarse [55] (pp. 87-88). Even in between one batch the microstructure changes according to the position of the sample on the casted piece [54]. Nevertheless, it is difficult to find a clear correlation between the overall microstructure and the mechanical properties. However, fatigue experiments on samples exhibiting different morphologies of the overall microstructure conducted by Rögner showed a quite uniform fatigue behavior [55] (p. 102).

2.5. Yttria stabilized zirconia

A detailed review on phase transformation in zirconia is given by Chevailer, including the following descriptions. Pure zirconium oxide appears in three different solid phases: monoclinic (m) at room temperature, tetragonal (t) above 1170 °C up to 2370 °C, and cubic (c) for higher temperatures. The (t-m)-transformation (from (t) to (m)) is martensitic and leads to a volume increase as well as to shear strains. The resulting internal stresses induce cracking during cooling down from the (t) temperature range. This can be prevented by adding oxides, which stabilize the cubic or the tetragonal phase at room temperature. Such oxides are typically calcium-, magnesium-, cerium- and last but not least yttrium-oxide. The latter one leads to zirconia ceramics showing the best combination of strength and toughness.

Depending on the amount of the stabilizing oxide, the tetragonal phase can be metastable below a certain temperature T_0. The (t-m)-phase transformation can then be triggered by stresses, e.g. by crack tip stress concentrations appearing during crack propagation (or by decreasing the temperature to the martensite start temperature T_{MS} [56]). The transformed zone around the crack tip induces shielding, accompanied with an increase in fracture toughness. The more the temperature falls below T_0, the lower the necessary stresses are to induce transformation. This results in a more pronounced toughening. For increasing Y_2O_3-contents, T_0 and hence the fracture toughness decreases. For very low Y-oxide contents, spontaneous (t-m)-transformation occurs during cooling. Therefore, a Y_2O_3-content of 2 mol % to 3 mol % is typically preferred. [7]

The toughness of tetragonal zirconia polycrystals (TZP) increases with increasing grain size [36, 56]. Swain et al. give an increase of T_{MS} related to an increasing grain size as a reason for enhanced toughening, according to the following mechanism. Cooling down from sintering temperature, internal residual stresses are induced by anisotropic thermal expansion and mismatch, mainly at grain corners, which can nucleate the phase transformation. These stresses can be relieved by diffusion down to a certain temperature, which is lower for smaller grains. Furthermore, the residual stress and strain rise with increasing grain size. The enhanced transformation for elevated grain sizes is equivalent to the formation of larger transformation zone sizes for coarser grains, appearing in front of a crack tip at a certain level of K_I. [56]

Higher applied stresses (i.e. stress intensities) also enhance toughening, since they cause a larger transformation zone. [7]

Transformation of metastable zirconia can also be caused by so called low temperature degradation (LTD, from room temperature to about 400 °C) in moist environment. However, at 37 °C it takes several years, before the monoclinic content of bulk 3Y-TZP samples can be expected to rise up to one or two percent according to diagrams shown by Chevalier. He suggests the following mechanism occurring during LTD: Ions such as OH^- diffuse via oxygen vacancies into the lattice and induce tensile stresses. Hence, (t-m)-transformation is nucleated at several surface sites and spreads gradually over the whole surface. Micro-cracks are formed and the transformation can expand to the bulk, because water penetrates into these cracks. [7]

2.5.1. Material properties

The Young's modulus for zirconia ceramics is 220 GPa according to Chevalier [36]. Liu et al. report a Young's modulus of 210 GPa and a Poisson's ratio of 0.33 for 3Y-TZP[39]. For 3Y-TZP with 0.38 µm grain size, a steep R-curve was found by Liu et al., similar to K_{RI} in figure 2.4 (Left), which rises during the first 20 µm of crack extension. They found this curve applying flaws of different sizes by indentation and conducting subsequent four-point-bending-tests in order to make the flaws grow [40].

Compared to zirconia ceramics such as Mg-PSZ and Ce-TZP, 3Y-TZP has limited transformation plasticity leading to a modest toughness and to a control of the high strength by the propagation of preexisting flaws [39]. Values for the fracture toughness reported in the literature are given in table 2.2.

Uniaxial fatigue tests on 3Y-TZP were conducted by Liu et al. (1 Hz, 1×10^5 cycles, $R = 0.8; 0.5; 0; -1$). They found fatigue failures to originate from internal preexisting flaws, but not to nucleate from the surface.

K_{IC} [MPa$\sqrt{m}$]	grain size [μm]	sintering	source
6.4	0.58	1500 °C, 2 h	[36]
7.0	1.07	1500 °C, 2 h and 1600 °C, 2 h	[36]
5.3	0.38	not available	[39]
4...5	0.38	not available	[40]
4.2...4.4	not available	1500 °C, 1 h	[57]
4.1	0.3	1450 °C, 1 h	[58] (pp. 29, 37, 67)
3...4	0.27	1400 °C, 1 h	[55] (p. 150)

Table 2.2. Fracture toughness values of 3Y-TZP found by various authors for different grain sizes and manufacturing parameters.

Fatigue fracture was always intergranular in their tests. Plastic strain accumulation and a reduction of stiffness was observed during fatigue, and was attributed to microcracking, which was identified as the dominating fatigue damage mechanism. A monoclinic phase content (determined using X-ray diffraction) of about 30 % was found on the fracture surfaces of the fatigued samples, which is the same content as it was found for monotonically investigated samples. [39]

Liu et al. also investigated the crack propagation of artificial flaws in bending experiments. These tests supported the idea that fatigue crack growth follows mechanisms related to microplasticity, more accurately microcracking. However, the latter cannot explain the steep crack growth exponent of 21. Due to similar crack growth exponents for Mg-PSZ (having a strong R-curve behavior) and for 3Y-TZP (showing a weak R-curve behavior), the authors conclude that this is not related to crack tip shielding but inherent to crack propagation kinetics in very brittle materials. [40]

In contrast, Chevalier et al. attribute cyclic degradation of 3Y-TZP to stress corrosion combined with a reduction of shielding mechanisms, such as bridging, phase transformation and microcracking. Switching between cyclic and static load with identic K_{Imax} showed a (about one order of magnitude) higher crack velocity for cycling. The transition between both velocities occurred always over a crack propagation distance of about 20 µm. Higher R-ratios which are equivalent to higher K_{Imax}-values or to smaller load amplitudes (see. equation 2.5) lead to a weaker reduction of shielding. According to Chevalier et. al. this can be explained by enhanced subcritical crack growth related to the higher K_{Imax}-value, which promotes reinforcement [e.g. by forming new bridges]. Furthermore, a reduced load amplitude means lower shielding degradation. For lower grain sizes of 3Y-TZP, the authors found a shift of the V-K_I-curve towards lower K_I-values. [36]

Pan [59] and Matsuzawa [60] found an anelastic response of 3Y-TZP in monotonic and in cyclic tests, mainly at lower stress levels (e.g. −200 MPa to 200 MPa). Anelastic strains were in the range of 2×10^{-5} to 6×10^{-5}. During cycling, anelasticity is suppressed, especially at higher frequencies such as 0.2 Hz. As possible reasons for the observed anelasticity, the authors suggest ferroelastic domain switching, slight shifts of oxygen ions, inverse phase transformation (monoclinic to tetragonal) and grain reorientation (the latter only in [59]). Tetragonal to monoclinic phase transformation was only found close to the strength [59]. Anelasticity is believed to relieve stress intensity at the crack tip and therefore work against crack propagation [60].

2.5.2. Micro low pressure injection molding

Micro molded ceramic parts can be manufactured by means of two different powder injection molding processes, namely high pressure powder injection molding (HPIM) and low pressure powder injection molding (LPIM, injection pressure 3 MPa to 5 MPa, working temperature ≤100 °C). Typically, arrays of micro parts form the part which is molded. For both processes, a ceramic powder (fine grained, e.g. 0.3 µm to 0.4 µm, for a low roughness of manufactured parts) is mixed with a polymer based binder, in order to obtain the feedstock, which is filled into the mold. Its viscosity must be low in order to achieve an adequate filling quality of microstructured details. For the same purpose, the mold is evacuated before filling. After solidification of the melt and demolding, thermal or catalytic debinding and subsequent sintering are required. LPIM green parts need a lower stability than HPIM green parts. Thus, binder systems with a lower viscosity are available for LPIM, enabling wall thicknesses in the range of 50 µm and aspect ratios larger than 10. Soft molds, e.g. made of silicon, can be used for LPIM in order to facilitate demolding. However, this is a manual process, which is only apt for small lot sizes, e.g. used for prototyping. [61]

The surfactant contained in the feedstock is an important parameter to optimize the surface quality and hence the strength of bending bars manufactured with LPIM. At the same time, edges tend to get more rounded and the geometry of the bars is likely to lose in accuracy, when the surfactant content is optimized to reduce surface roughness. Taking into account the change in geometry of the bar cross section due to the rounded edges, the bending strength rises compared to rectangular cross sections. [62]

2.6. Small scale fatigue testing methods

In a review on fatigue testing of polysilicon, Sharpe and Bagdahn [63] presented several MEMS-based (Micro Electro Mechanical System) testing techniques for samples having thicknesses in the range of 1 µm to 30 µm and widths in the range of 4 µm to 200 µm. They are manufactured together with the actuating system on a substrate from which movable parts (including the specimen) are typically released by etching. The working principles of the actuators are based on electro statics, electric heating, electro magnetics and on the inverse piezo effect. The mentioned frequencies are 1 Hz (electric heating) and 20 as well as 50 kHz. Most of the systems work in resonance and apply bending on a notched sample or on a cantilever beam. The crack length can often be correlated to the change in resonance frequency. The setup based on electric heating works in tension-tension fatigue.

In the same review, the authors present one setup they built for tensile-tensile fatigue. In this case the samples are again manufactured on a substrate and released by etching but transfered into the fatigue setup which is not MEMS-based. However, it consists of a load cell on a three axis stage and a piezo or a voice coil actuator controlled by a personal computer. The authors conducted fatigue tests at 50 Hz with a peak load of less than 0.2 N. The design of one of the setups built in the present work (see. chapter 3) is based on this setup design from Sharpe. [63]

Numerous studies on microscale silicon were motivated by this question: why does silicon show fatigue behavior at the microscale but not at the macroscale? Alsem et al. attributed this effect to moisture-induced subcritical crack growth inside an oxide layer on the silicon surface, the thickness of which increases due to cyclic stresses occurring during fatigue [64]. Other researchers like Kahn et al. favored mechanically induced subcritical cracking in the silicon as an explanation for fatigue behavior [65].

The various testing techniques and test conditions which were used to reveal the fatigue behavior of microscale polycrystalline and singlecrystalline silicon were reviewed by Alsem et al. [64]. The authors distinguish between MEMS-devices containing the actuator as well as the sample and other devices, which are driven externally. External actuation was realized with an atomic force microscope (AFM), piezoelectric actuators, voice-coils, electromagnetic devices or nanoindentation testers. On-chip actuators were mainly based on electro-static principles, but also piezoelectric, electromagnetic and thermal actuations were realized. Sample thicknesses ranged from 0.255 µm to 30 µm. The majority of the reviewed studies were conducted at frequencies in the range of about 10 kHz to 50 kHz and at load ratios of $R = -1$. However, frequencies between 0.1 Hz and 1000 Hz as well as experiments at 13 MHz are also reported. Furthermore, load ratios of $R = 0.1$ and $R = 0$ were applied [64]. Kahn et al. were able to vary the load ratio between $R = 0.5$ and $R = -3$ with their testing device [65].

Boyce et al. investigated 26 µm thick Ni-cantilevers in fully reversed bending, which were manufactured by electrodeposition (LIGA-process). For this purpose, they used a displacement controlled custom built servohydraulic device with a load cell. The experiments were conducted at 20 Hz and revealed broken surface oxides at the location of persistent slip bands as a key mechanism for fatigue of the investigated Ni-samples. The broken oxide enables diffusion of atmospheric oxygen and further oxidation of underlaying metal. [66]

Two experimental techniques to investigate fatigue of metallic thin films deposited on elastically deforming substrates were presented by Kraft et al. [67]. First, cyclic tensile tests of 3 µm thick Cu films deposited on a polyimide substrate were driven using an electromechanical device. The test is load controlled (dominated by substrate), and the unknown stress-strain behavior of the film is determined using X-ray diffraction. Second, beams with 0.2 µm to 1.5 µm thick Ag-films on a 2.8 µm thick SiO_2 substrate are cyclically bent at 45 Hz by means of a nanoindenter in load control. The resulting strain can be calculated using the elastic beam theory, the sample geometry and the measured load. Since the strain leads to yielding of the film, the stress cannot be calculated. [67]

Thin films were fatigued at frequencies in the 900 MHz-regime up to several 10^{12} cycles by Eberl et al. [68] using surface acoustic wave filters (SAW). A special metallic circuit path pattern (lines out of Al, 420 nm thick, 1 µm wide) is deposited on a piezoelectric substrate. Applying an AC signal at the resonant frequency of the device, a standing surface acoustic shear wave is created. Thus, the metalization is loaded cyclically with a maximum shear stress appearing at the interface between substrate and metalization. Due to fatigue, voids and extrusions are formed in the metalization, inducing a change of the SAW's natural frequency. Thus, damage can be detected observing the change in natural frequency. [68]

In situ fatigue experiments on Cu wires having 30 µm to 50 µm diameters were conducted in a scanning electron microscope by Dehm et al. at frequencies up to 10 Hz. Cyclic motion is induced by a piezo transducer. Motor driven, or piezo driven stages are used to maintain the maximum load during the experiment, to position the sample and to clamp it using tweezers. The load is measured with a scatter of ±3 mN, detecting the frequency change of an oscillating wire, which is kept in tension inside a magnetic gap and connected to a flexible stage carrying the tweezers. [69] [70]

Kiener et al. conducted cyclic in situ cantilever bending experiments on single crystalline copper in a SEM. The samples (cross sections $9.5 \times 10.5\,\mu m^2$ and $2.2 \times 9\,\mu m^2$) were cut out using a focused ion beam (FIB) and bent by means of a micro indenter. Up to 100 displacement controlled cycles were applied (load ratio $R = -1$, 0.033 Hz), and SEM images were taken after distinct cycles, allowing for the detection of arising slip lines due to local strain redistribution. [71]

Up to 47 load cycles were applied on electron transparent nanotwinned copper films (thickness about 100 nm) by Funk in a transmission electron microscope (TEM), in order to observe micro-structural changes. The samples were hourglass shaped having a gage width of about 1 µm and were cut out of a 20 µm thick sputter deposited film using a FIB. They are glued on a micro mechanical push-to-pull device using platinum. A picoindenter mounted into a TEM pushes on this device, which converts the motion into traction of the sample. The loading is displacement controlled and strain is measured optically using digital image correlation. [72]

Khatibi et al. fatigued ultrafine grained Cu samples with a gage of 1500 µm length (cross section $600 \times 200\,\mu m^2$, notched in the center) at a frequency of 20 kHz. For this purpose they adapted a standard ultrasonic VHCF-testing device (such as mentioned above, [16]). They removed material from the resonating setup part at the point of maximum push-pull strain amplitude by drilling a hole. The tested micro sample was glued at this location (over the hole, thus the sample gage section is freestanding) in order to load it cyclically. A strain gage was applied on the resonating part in order to deliver the signal for a strain controlled experiment which is interpreted as being stress controlled assuming elastic deformation and a theoretical notch factor. It was calibrated using a dummy sample carrying another strain gage. The same technique was also used to test Cu micro wires with diameters from 10 µm to 125 µm. [73] [74]

3. Custom built setups for mechanical micro sample testing

In the last section of the previous chapter, several small scale fatigue testing techniques found in the literature were presented. These techniques are typically designed for experiments on a special type of sample conducted at a certain frequency. One aim of the present work was to extend the pool of available testing methods in order to gain flexibility in terms of sample geometries, materials, and frequency ranges accessible by experiments. Another goal was to keep the cost of the setup design at a reasonable level. For this purpose, custom built setups were developed, based on the following components: A piezo actuator, two x-y stages, a load cell or capacitive displacement sensor (see below), a machine vision camera, a microscope and a personal computer with a data acquisition board (providing analog inputs and analog outputs) driven by a LabView programming environment. Different setup-configurations that were developed based on these components are presented in this chapter.

3.1. Setups for monotonic and cyclic testing at up to 100 Hz

The custom built flexible micro sample tester shown in figure 3.1 can be used for monotonic tests such as quasi static tension, compression or bending as well as for fatigue tests up to 100 Hz. In these tests the sample is mounted between a fixed part and a moving part of the machine in order to load it mechanically. A piezo actuator induces a linear motion along the x-axis (closed loop resolution: 1.2 nm). It carries a dynamic load cell which is attached to the moving side of the test specimen. Therefore, the load acting on the sample can be measured. The fixed side of the sample as well as the piezo actuator (moving side of the sample) are mounted onto an x-y-stage. Hereby, they can be positioned transversally to the loading direction while the sample is set up in order to prevent constraints on the sample due to misalignment. The limited travel range of the piezo actuator can be enlarged by moving the stage in the loading direction x. For the purpose of aligning the sample and of measuring strain during the experiment, the sample surface is observed by means of a CCD camera combined with an optical microscope.

The whole setup is installed on an optical table in order to isolate it from vibrations coming from the environment. Such vibrations would affect the sensitive load and strain measurement. In an early state of the setup development, all stage movements were induced manually by means of micrometer screws. A refined version contains a stepper motor, which drives the stage in the x-direction for enlargement of the piezo actuator's travel range. The stepper motor pushes against the stage without being connected to it and only a weak spring inside the stage assures contact between motor and stage. Therefore, a small load pulling the stage away from the motor would

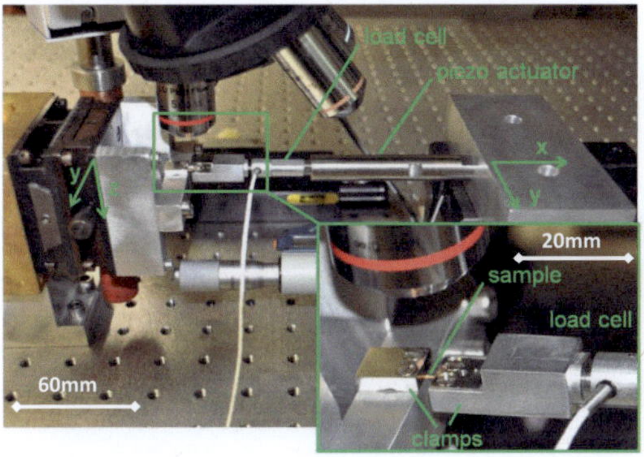

Figure 3.1. Custom built setup for monotonic tests and for fatigue tests (setup 1).

lead to a loss of contact. A counter weight with a deflection roller on the table was installed, which pulls the stage always against the motor. The weight is damped in an oil bath in order to prevent vibrations.

device	range		resolution		stiffness		company
load cell	±50	N	25	mN	45	$N\,\mu m^{-1}$	FGP Sensors, France
piezo actuator setup 1	60 +300 −800	µm N N	1.2	nm	15	$N\,\mu m^{-1}$	Physik Instrumente, Germany
motor setup 1	25 ±70 0.8	mm N $mm\,s^{-1}$	0.1	µm	not specified		Physik Instrumente, Germany
motor setup 2	25 ±20 20	mm N $mm\,s^{-1}$	0.2	µm	0.25	$N\,\mu m^{-1}$	Physik Instrumente, Germany

Table 3.1. Characteristics of the mechanical devices contained in the setups 1 and 2.

Two further setups are presented in this section, the development of which was part of this work. A simple low cost version of the setup described before is shown in figure 3.2. The load cell with grip is mounted onto a stage which can be driven manually in y- and in z-direction for the purpose of alignment. A servomotor moves along the x-axis in order to apply the load on the sample. The resolution for positioning, the stiffness and the maximum achievable load are limited compared to setup 1 (see table 3.1). In order to realize a smooth motion at low velocities, the PID-parameters of the servo motor had to be adjusted. However, the quality in smoothness for small velocities is still limited compared to the piezo-motion of setup 1.

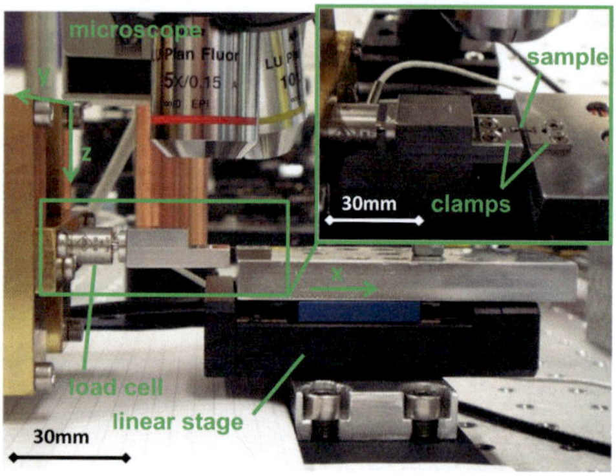

Figure 3.2. Custom built setup for monotonic tensile tests driven by a servo motor (setup 2).

A voice coil driven fatigue setup which was developed at the Johns Hopkins University of Baltimore, USA by Prof. Chung-Youb Kim and Prof. W. N. Sharpe, Jr. is presented in figure 3.3. A small contribution to the development of the setup was part of this work: the laser system for the strain measurement was designed and first measurements were done to validate the system. Valuable experiences made during this cooperation helped to develop the setups presented in this work.

A voice coil applies a sinusoidal tensile load on the sample which is measured by means of a dynamic load cell. Strain is measured optically. For this purpose, laser spots are pointed to the front and the back side of an hourglass shaped micro sample equipped with reflective markers on each surface. Measuring strain on both sides of the sample enables detection of bending due to misalignment. On each sample side two fringe patterns ap-

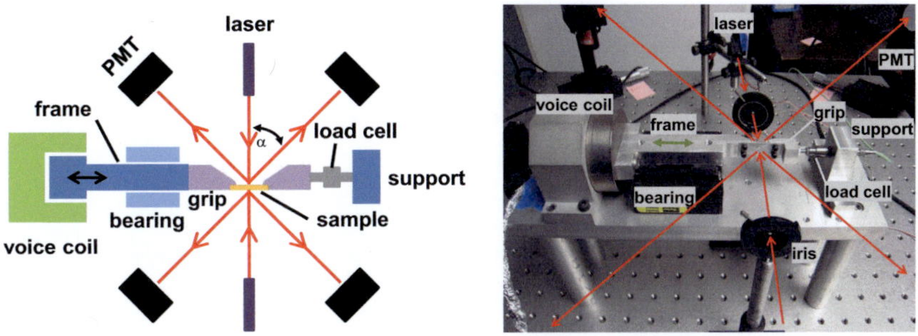

Figure 3.3. (Left) Schematic of a custom built fatigue setup developed at the Johns Hopkins University in Baltimore (first tests at 27 Hz); (Right) Picture of the setup: red lines indicate the path of the laser beam.

pear at an angle α to the incident laser beam. The intensity of each pattern is detected by Photo Multiplier Tubes (PMTs) for strain measurement. The strain values found with the two PMTs per side are averaged to eliminate measurement errors due to rigid body motion (see [75]). Further details about the theory of this measuring method are given in section 3.1.1. Since the voice coil actuator has an elevated inertia, a precise control is difficult for high frequencies. The experiments were conducted at 27 Hz. Gripping of the sample was based on the principle shown in figure 3.5 (Left) (see p. 66).

The PMTs are mounted into black painted metallic boxes in order to protect them from ambient light as they can easily get damaged when receiving too much light intensity during operation. A small rotary plastic cap covering a hole in the metallic box has a narrow slit as light inlet. During strain measurement the slit is directed to a fringe of the interferometric pattern, so that a sufficiently large intensity change is detected when the fringe moves away from the slit. The slit width is slightly less than the width of a fringe. In order to sweep through the fringe pattern during calibration fringe by fringe, each PMT sits on a linear stage. In fact, for small strain signals, intensity maxima and minima of the fringe patterns must be determined before the fatigue experiment, because they are needed to calculate the relative intensity change which is related to the applied strain (see 3.1.1). The control of the voice coil and the data acquisition (load cell and PMT-signals) are conducted by means of a LabView program. Strain calculation of the program version used for the first experiments was not completely automatic. An intelligent algorithm recognizing reliably intensity maxima and minima without interaction of the user was not established but is necessary for a reliable strain calculation. One challenge is to distinguish automatically between intensity changes in the detected fringe pattern due to loading and unloading of the sample.

3.1.1. Strain measurement

Two methods of optical strain measurement will be presented in this section. First, the laser based Interferometric Strain/Displacement Gage (ISDG) applied in the voice coil driven setup presented above will be explained. Second, the Digital Image Correlation (DIC) based on images taken with a digital camera will be introduced. This method was used for strain measurements during quasi-static experiments.

The ISDG is a very fast (20 kHz) and precise (strain resolution of 1×10^{-5}) technique to measure strain. Two reflective markers at an appropriate distance d are applied onto the surface of the specimen and illuminated with a laser. In analogy to an optical double slit experiment, two fringe patterns depending on the distance d between the markers (gage length) appear [76]. The basic principle is explained in the following according to [77]:

One can observe a fringe (intensity maximum) of the order m at an angle α relative to the incident laser beam. The following equation is valid:

$$d \sin \alpha = m\lambda \tag{3.1}$$

Herein λ is the laser's wave length. Dilatation of the sample induces a change of d which leads to a change of α. The applied strain ϵ is related to the relative position change of the fringe ${}_{\Delta}m$, initially located at the angle α_0.

$$\epsilon = \frac{{}_{\Delta}m\lambda}{d_0 \sin \alpha_0} \tag{3.2}$$

In this relation d_0 is the distance between the markers on the unloaded sample. When a fringe moves to the initial position of its neighboring fringe, ${}_{\Delta}m$ equals 1. The intensity I of the fringe pattern is known (see equation 3.3).

$$I = I_0 \cos^2 \frac{\pi d \sin \alpha}{\lambda} \tag{3.3}$$

One can measure this intensity at a fixed observation position α using photo multiplier tubes (PMTs). This approach is adapted for measuring strain at high frequencies. A movement of fringes is related to a change of the detected intensity. When a fringe passes at the observation point, the maximum intensity I_0 is measured. Inverting the above equation, the distance between the markers d_i can be related to the intensity as follows.

$$d_i = \frac{\lambda \arccos\left(\sqrt{I/I_0}\right)}{\pi \sin \alpha} \tag{3.4}$$

The index i was introduced for d_i because the value corresponds to a change in marker distance between intensity peaks and valleys and not to an absolute value of d as will be explained in the following. Finally, the strain

ϵ can be calculated using the current distance d and the initial distance d_0 (unloaded state) between the markers on the sample.

$$\epsilon = \frac{d - d_0}{d_0} \tag{3.5}$$

According to equation 3.3, the measured signal gets periodically the same even when the distance d rises continuously. Therefore the inverse function of equation 3.4 is only valid for a small range of d_i without additional data treatment. As soon as an intensity maximum ($I = I_0$) or minimum ($I \approx 0$, the measured value I is always a number greater than 0) is reached, the change of the term I/I_0 changes in sign, and so the change of d_i does, according to equation 3.4. Additional information from the experiment (such as the loading direction) is necessary, in order to know, whether the intensity change is due to forward (loading) or backward motion (unloading) of the fringes. Depending on this information d_i must be added to or subtracted from the value d reached at the last intensity maximum or minimum.

Using ISDG, the strain is measured locally. A typical marker distance d, which is the equivalent to the gage length for strain measurement, is 200 µm. Deposited platinum or gold lines as well as indents on the sample surface can serve as "markers". The sample surface must be polished in order to obtain a good intensity signal. Indentations and platinum lines were tested on as-cast aluminum bronze samples, but the needed fringe pattern could not be observed. The sample roughness causes a diffuse reflection of the light which covers the fringe pattern.

The second method of optical strain measurement is digital image correlation (DIC). This method is based on the analysis of images made from the sample during experiments. Functions of the software Matlab, which are online available, are used to determine strain with the presented setups

[78]. Subsections of the images (typically at a size of about 30 × 30 pixel) are defined as markers and tracked from image to image. When the sample is displaced or deformed from image i to image $i+1$, the markers move. The new marker positions in the image $i+1$ can be identified by means of an image correlation algorithm. Deriving the marker displacement between two images gives the strain of the sample in the region of the markers. The method is very convenient for post processing, once the images are taken during the experiments. Markers can be set to points of interest on the sample and correlation parameters can be optimized in order to get the best result in terms of accuracy. However, compared to the ISDG, the DIC is limited in speed. The achievable frame rate of the employed digital camera determines the maximum speed of strain measurement. For live measurements during the experiment (in contrast to post processing), the speed of the correlation algorithm is an important limit. The fewer markers are defined on the image and the smaller their size is, the faster the algorithm works. New approaches running a DIC-algorithm on the graphics board of the computer enable fast data treatment, and deliver strain values at a speed which corresponds to a live view for quasi static tests.

Calculating the strain from the displacement is kept as simple as possible. The camera is aligned with respect to the loading direction in a way that either the rows (x-direction of the picture) or the columns (y-direction of the picture) of pixels coincide with this direction. Thus, the x- and y-coordinates of markers calculated by the correlation algorithm correspond to marker positions in the loading direction and transversal to it. In the following, each of the M markers set on the first image are numbered by an index k ($k = 1, 2, ..., M$). For each image having the number i ($i = 0, 1, 2, ..., N$) the displacement for each marker u_k in x-direction and v_k in y-direction on the picture i is calculated using the coordinates x_k and y_k of the marker position on picture i as follows:

$$u_k^i = x_k^i - x_k^0 \tag{3.6}$$

$$v_k^i = y_k^i - y_k^0 \tag{3.7}$$

The strain is calculated using the marker displacement with the following relations which are valid for small deformations ($|\epsilon| \ll 1$)

$$\epsilon_x = \frac{\partial u}{\partial x} \tag{3.8}$$

$$\epsilon_y = \frac{\partial v}{\partial y} \tag{3.9}$$

For this purpose, the displacements u_k^i (or v_k^i) from all k markers are plotted over the marker position x_k^i (or y_k^i). This corresponds to the Eulerian description where the displacement is a function of the deformed system. To describe the mechanics of elastic structures it is more common to use the Lagrangian description, where the displacement is expressed as a function of the non deformed structure [79] (p. 92). This could be done by plotting the displacement u_k^i over the marker position of the first image x_k^0 showing the undeformed sample. For small deformations the difference between the two descriptions is negligible. Only markers from a homogeneously deforming region are selected. Thus, all data points of the displacement-position-plot lie on a straight line. The slope of a linear least square fit to the points equals the strain of the homogeneously deforming sample section, corresponding to equations 3.8 and 3.9.

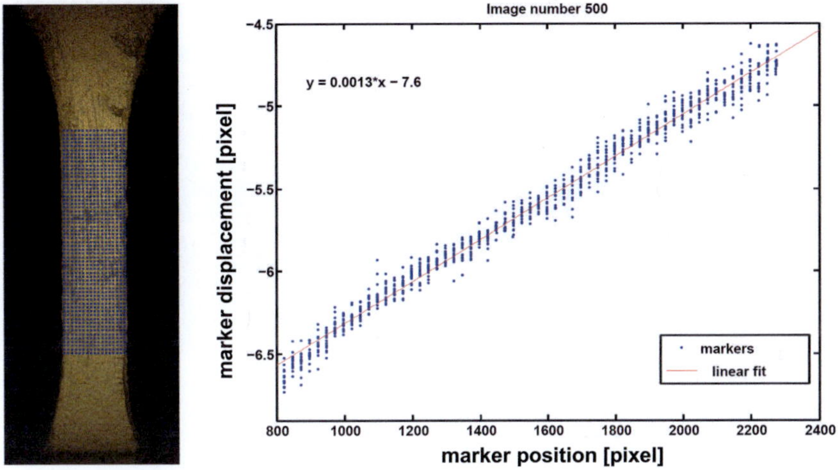

Figure 3.4. (Left) Marker grid on sample; (Right) Displacement over marker position (2 pixel = 1 µm).

An image of a tensile test on yttrium stabilized zirconia is shown with an overlayed marker grid for the digital image correlation in figure 3.4 (Left). The displacement of the markers on the deformed sample in tensile direction is plotted over the marker position in figure 3.4 (Right). The linear fit gives a strain of 1.3×10^{-3} after equation 3.8.

3.1.2. Micro tensile tests

Two different kinds of sample grips were developed for tensile tests on setup 1 (figure 3.1). The first version is designed for tensile samples having triangular clamping sections (see figure 3.5 (Left)), similar to the self-aligning grips shown by [80]. Before testing, the grips are moved quite close together so that the sample can be placed easily into the grips. When the grips are moving apart, the load on the sample increases as soon as the length of the sample is reached. Thus, the latter is aligned automatically in the tensile direction by means of the triangular shaped clamping sections.

Figure 3.5. (Left) Tensile sample (nanoporous gold) having triangular shaped clamping sections in self-aligning grips; (Right) Ceramic sample having rectangular clamping sections, which are covered by the grips and not visible in the image.

When the micro tensile samples have the conventional flat dog bone geometry, they must be aligned optically. For this purpose, the clamp sitting on the actuated part of the setup (see figure 3.1) is demounted and placed on a table. Using a magnifying glass, the sample is aligned with respect to the edge of the grip and one end is clamped between the grip and a sheet of steel, which is fixed with two screws. Afterwards the grip carrying the sample is mounted onto the load cell, which is then fixed to the piezo actuator. Looking through the microscope, the piezo actuator can be rotated until the sample is adjusted rotationally. Now the piezo is tightened and the motor moves (along the x-axis, figure 3.1) until the free sample side is positioned above the cross beam (clamp, which does not move during the experiment). The piezo is moved out as far as possible to provide the travel range for the tensile experiment. The cross beam is moved upwards until it touches the sample (visible in the microscope). A metallic sheet is again screwed on top of the sample's gripping surface in order to attach it to the cross bar. For brittle samples, such as ceramics, super glue is applied between the grips and the sample. Thus, the screws can be tightened just lightly in order to avoid fracture of the sample in the griping area. The elastic glue layer homogenizes the contact pressure between grip and sample. A ceramic sample, which was clamped this way and observed through the microscope, is shown in 3.5 (Right).

Load measurement is started before the mounting procedure. However, usually a new load file with higher sample rate is started just before the experiment. Also, the image acquisition of the camera is started, mostly at a rate of 1 frame per second, in order to obtain images from the sample gage for the strain measurement by digital image correlation. Afterwards, the piezo motion is started at the desired speed. For ductile samples it can be the case that the travel range of the piezo is not sufficient. In this case, the piezo is switched to the load controlled mode. The motor can be moved slowly in compression direction of the sample, which is compensated by the piezo maintaining the same load. When the piezo is moved out completely, the motor is stopped, and the piezo travel range is again available in tensile direction.

With this setup configuration the following micro samples were tested (cross section in parentheses): Aluminum bronze (260 × 130 μm^2, part of this work), nanoporous gold (260..500 × 220 μm^2) [81], nanotwinned copper (260..500 × 20 μm^2) [72], yttria stabilized zirconia (210 × 120 μm^2, part of this work), Aluminum (260 × 200 and 500 × 200 μm^2, to develop the setup and the self-aligning grips).

3.1.3. Micro compression tests

Compression tests on ceramic pebbles (Li_4SiO_4) with a diameter of 500 µm were conducted using an intermediate setup design of the development procedure. The configuration is shown in figure 3.6. A fix plate and the piezo actuator, which carries a load cell, are aligned on an optical rail. The ceramic pebble is attached at the fix plate with the residual glue from a piece of tape. The piezo actuator is driven at constant velocity towards the pebble until it crushes. By means of the load cell, the crush load can be determined. Using image correlation, a load displacement curve can be derived. The development of the setup was part of this work. Experiments were conducted by Zhao [82, 83]. With the same setup configuration also cyclic indentations were performed. Further compression tests were conducted on a polymer metamaterial using setup 1 by Bückmann [84]. These applications are mentioned in order to emphasize the flexibility of the setup components.

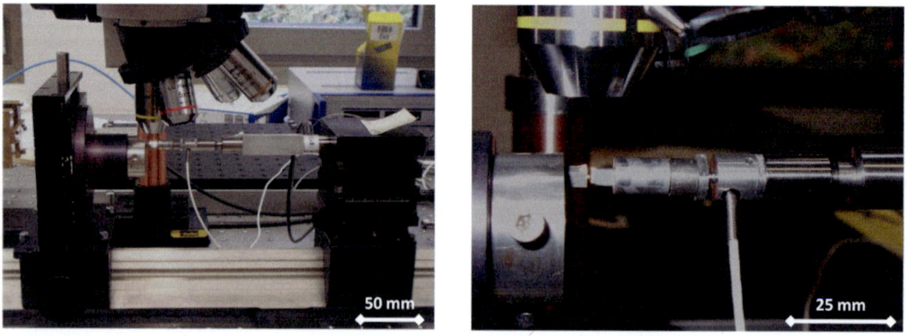

Figure 3.6. (Left) Custom built setup for compression tests; (Right) Detail of the setup: a ceramics pebble is compressed.

3.1.4. Micro bending tests

Three-point-bending tests were conducted with the setup 1 (see figure 3.1). For this purpose, the clamps were replaced by custom built bearings for three-point-bending (see figure 3.7). The shafts of drills with a diameter of 290 µm were used as rolls in contact with the sample. The center role consists of half a shaft glued onto the tapered tip of a ceramic rod (fixed support in figure 3.7 (Right)). The latter is attached at the x-y-stage of setup 1 (see figure 3.1 left border). The two outer rolls (moving support in figure 3.7 (Right)) are mounted onto a steel block carried by the piezo actuator. They are guided by slits implemented in the steel block and lie on a ceramic inlet glued into the steel block. Therefore they are free to rotate (neglecting friction between the rolls and the guiding slits) in order to minimize friction during the experiment [85].

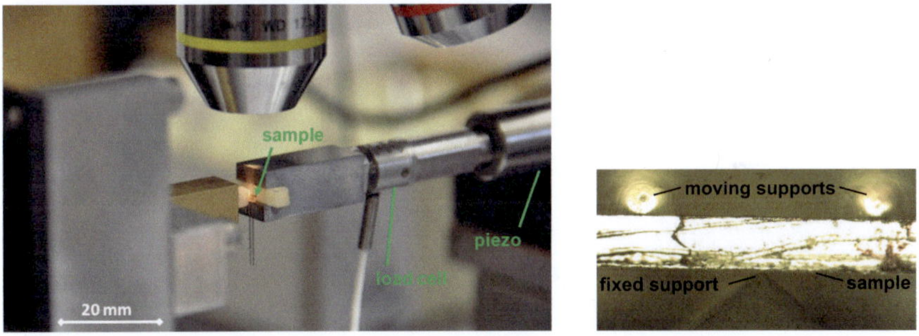

Figure 3.7. (Left) Setup from 3.1 adjusted for bending experiments; (Right) Picture of a three-point-bending sample observed through the microscope during a bending experiment.

In order to mount the sample, the x-y-stage of setup 1 which carries the piezo actuator (right border in figure 3.1) is driven away from the fixed support. The sample is positioned on the moving supports using tweezers. Since the setup is arranged horizontally, the samples must be prevented from falling down. For this purpose, a small drop of water is applied before.

The adhesion is enough to keep the sample in position until contact with the fixed support is established.

The fracture toughness of nanocrystalline nickel was investigated with this configuration [85]. For the present work, fatigue tests on yttria stabilized zirconia were conducted. In this case, it was taken care that the water on the sample evaporated before starting the test, since water molecules are known to enhance subcritical crack growth, i.e. damage evolution during fatigue [36].

3.2. Resonant micro fatigue setup

The setups presented above were not used for the very high cycle regime. Applying a very high number of cycles at considerable load on the load cell for each fatigue test is expected to affect the reliability of the load cell. Furthermore, the load cell with clamp causes a natural frequency of the setup which defines an upper frequency limit for fatigue tests. Therefore, a resonant fatigue setup working without a load cell was developed and is presented in this section. One version of the setup is shown in figure 3.8. A piezo actuator is fixed to a frame sitting on an optical table (to eliminate vibrations coming from the environment). The actuator's main axis is aligned in the direction of gravity. One end of the sample is attached to the actuator. A mass is clamped onto the other end of the sample. The mass moves up and down sinusoidally when excited by the piezo actuator. Thus, a dynamic load is induced on the sample. The mass displacement is measured without contact by means of a capacitive displacement sensor. Therefore, the inertial load acting on the sample can be determined.

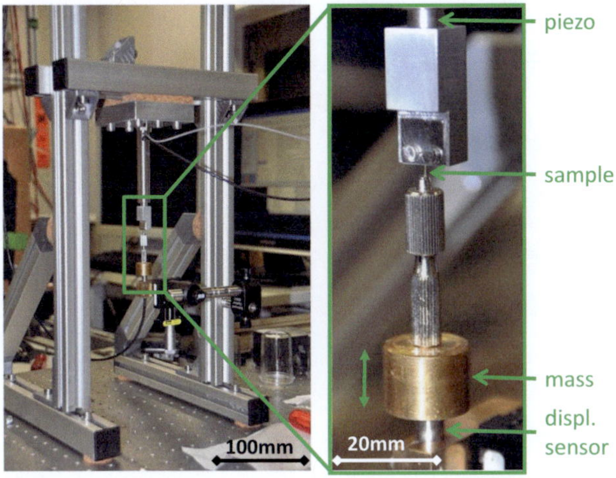

Figure 3.8. Custom built resonant setup for fatigue in the kHz-regime.

3.2.1. Theoretical approach

A simplified mechanical model of the setup is shown in figure 3.9. Herein, the sample and the mass are represented by a mass spring oscillator, which is excited by a perfectly stiff actuator. In reality, the system is more complex since the piezo actuator has a limited stiffness and complies when a mechanical load is applied. Moreover, piezo actuators show complex behavior since mechanical and electrical fields are coupled. A load applied on the piezo actuator induces an electrical field and vice versa. However, the actuator properties can be estimated by a linear mass spring behavior, in particular for small signal conditions, i.e. small loads [86] (p. 2-198). Together with the mass spring oscillator in figure 3.9, this would lead to two connected mass spring oscillators (which is a mass spring oscillator with two degrees of freedom). In addition, damping in the piezo actuator system (caused by friction and by electric resistivity) as well as damping in the sample system (mass spring oscillator from figure 3.9, caused by the aerodynamic resistance of the mass) exist in reality. However, the simplified mechanical model can be used to determine the load acting on the sample, as will be explained in the following.

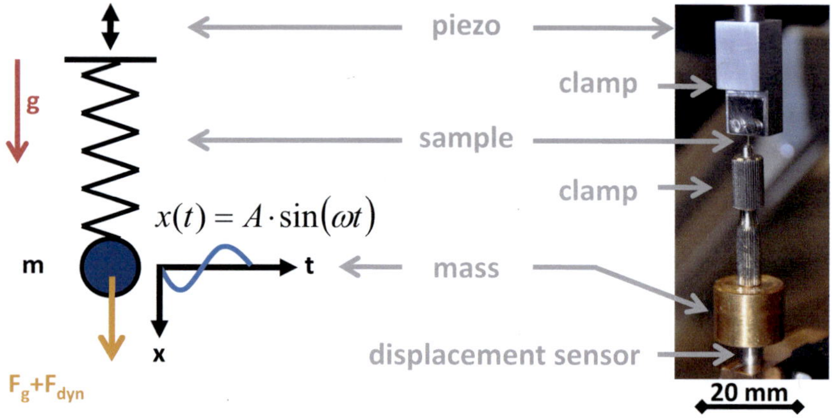

Figure 3.9. Mechanical model of the resonant setup.

When the lower half of the sample is cut free together with the mass and all external loads, dynamic loads and dead loads are considered, the internal force of the sample can be calculated using the balance of all forces. The dead load is equal to the weight of the mass. The only external load is caused by air resistance. Dynamic load and air resistance can be calculated when the motion (that is to say velocity and acceleration) of the mass is known. This consideration is independent from the complexity of the actuating system. The interaction of a complex actuating system with the mass spring oscillator could lead to a complex motion of the mass. Nevertheless, the load acting on the sample can be calculated when the motion of the mass is known. The experiments conducted with the presented setup showed that the sample moves sinusoidally, which makes the simplified model applicable.

Let us come back to the mechanical model of figure 3.9 with a perfectly stiff actuator. The applied loads lead to a deformation of the sample which is (macroscopically) elastic (High Cycle and Very High Cycle Fatigue), so that we can represent the sample as a linear elastic spring in the mechanical model. The other components (e.g. clamps), designed in a much stiffer way than the sample, are considered to be rigid bodies. The mass and the clamp, which are attached to the lower end of the sample, are represented as a single point mass. The mass of the sample can be neglected since it is orders of magnitudes smaller.

The system is excited by a sinusoidal motion at the upper end of the sample close to its natural frequency f_0. The latter is determined by the stiffness c of the sample and by the mass m fixed at the bottom of the sample according to equation 3.10.

$$\omega_0 = 2\pi f_0 = \sqrt{c/m} \tag{3.10}$$

Herein ω_0 is the natural angular frequency. This leads to a sinusoidal displacement x of the mass at the lower end of the sample, which varies over time t as follows.

$$x(t) = A \sin(\omega t) \tag{3.11}$$

In this equation, A is the displacement amplitude of the mass and ω is the angular frequency at which the mass vibrates. In order to calculate inertial forces induced by this motion, the acceleration acting on the mass m is calculated differentiating equation 3.11 two times with respect to x.

$$\ddot{x}(t) = -A\omega^2 \sin(\omega t) \tag{3.12}$$

Multiplying equation 3.12 by m gives rise to the inertial force which is equivalent to the dynamic load acting on the sample. Adding the weight of the mass (mg), which is the superimposed static load, we obtain the total load $F(t)$ acting on the sample (see equation 3.13).

$$F(t) = mg + mA\omega^2 \sin(\omega t) \tag{3.13}$$

Herein, g is the gravity acceleration. Consequently, $mA\omega^2$ is the dynamic load amplitude applied during the fatigue test. It can be adjusted by controlling the displacement amplitude A of the mass.

The amount of the mass is chosen in a way that the natural frequency lies in the range of the desired fatigue frequency (see equation 3.10), which is one or two kHz in our case. Since the system is excited at its natural frequency, the static load induced by the mass is related to the excitation frequency and cannot be chosen freely. The same is valid for the load ratio R (see equation 3.14).

$$R = \frac{\sigma_{min}}{\sigma_{max}} = \frac{F_{min}}{F_{max}} \tag{3.14}$$

σ_{min} and σ_{max} are the minimum and maximum stresses generated in the sample during cyclic loading. Since for the present solicitation the stress is calculated dividing the load by the cross section of the sample, R can also be calculated by the quotient of minimum and maximum load. Combining equations 3.13 and 3.14, R can be expressed by

$$R = \frac{m(g + A\omega^2)}{m(g - A\omega^2)} = \frac{g + A\omega^2}{g - A\omega^2} \tag{3.15}$$

As a result, changing the load amplitude implies a change of the load ratio. In our application the static load is small compared to the dynamic load amplitudes so that R varies only in between -0.9 and -1.

The formation of damage during the fatigue test (e.g. formation of cracks) leads to a decrease of the sample stiffness c. Equation 3.10 shows that this causes a decrease in natural frequency of the system. If the excitation frequency is not adapted, a remarkably higher excitation amplitude is necessary to induce the desired displacement amplitude of the mass. This can be used to detect fatigue damage of the sample.

The accuracy of clamping has an impact on the sample stiffness which is not negligible due to the small sample geometry (0.2 mm change in clamping position corresponds to about 6 % stiffness change which induces a change in frequency of about 3 %). For this reason and due to the influence of the manufacturing process on the sample stiffness (cavities and imperfection in geometry) the natural frequency is different for different experiments.

In order to simplify the analysis, damping was neglected. The only external load acting on the mass and thus inducing damping is the air resistance, which depends on the velocity of the mass. The resulting force of drag F_d which acts in the opposite direction of the mass's motion can be estimated using the following equation [19] (p. B59).

$$F_d = \frac{1}{2} c_d \rho v_0^2 A_p \tag{3.16}$$

Herein c_d is the drag coefficient, ρ is the density of the air, v_0 is the relative speed between air and mass and A_p is the projected area of the mass (on a plane perpendicular to the direction of motion). Using typical values from the experiments conducted in this work, it can be seen in the following example that F_d is very small compared to the load applied on the sample:

c_d: 1.11 for a cylindric plate [19] (p. B60) (overestimated, for the cylindric mass c_d is lower)
ρ: $1.20\,\mathrm{kg\,m^{-3}}$ (air at 20 °C)
v_0: $35\,\mathrm{mm\,s^{-1}}$ (maximum speed during a sinusoidal motion at 1000 Hz with an amplitude of 5.5 µm)
A_p: $255\,\mathrm{mm^2}$ (surface area of the cylindric mass with 18 mm diameter)

With these values we obtain a maximum force of drag F_d acting on the mass of $2.3 \times 10^{-7}\,\mathrm{N}$. This value is negligible compared to the load amplitude of 7.4 N, which is applied on the sample. Since F_d is proportional to the square of v_0 it is zero at the reversal points of the mass where the sample load is maximum.

Because the free hanging mass is not supported laterally, no frictional force between moving parts such as a bearing must be considered. The excitation close to the natural frequency of the sample's push-pull mode ensures that the solicitation is applied along the desired direction. However, we have to deal with a free hanging mass that can swing transversally like a pendulum. In order to avoid such a motion as far as possible the setup is mechanically isolated from the environment. This isolation is not perfect and the manipulation during the mounting procedure also implies small excitations. In addition, a small portion of the excitation energy coming from the piezo can be dissipated in other motions than the excited Eigenmode, due to imperfections of the real system. As a result, a small pendulum motion can be superposed on the desired push-pull motion. In our case the frequency of the pendulum is in the order of 2 Hz far below the fatigue test frequency of about 1000 Hz. For this reason and due to the limited amplitude of the transversal motion cyclic damage induced by pendulum motion can be neglected.

Using equation 3.13 divided by the sample cross section ($t_s \times w_s$) one can estimate the error in measuring the applied maximum stress (i.e. $\sin(\omega t) = 1$). For this purpose the derivatives with respect to the relevant input parameters must be calculated:

$$\frac{\partial \sigma}{\partial t_s} = -\frac{mg + mA\omega^2}{w_s t_s^2} \tag{3.17}$$

Herein t_s and w_s are the sample thickness and width respectively. The derivative with respect to w_s is the same equation, when w_s and t_s are interchanged.

$$\frac{\partial \sigma}{\partial A} = \frac{m\omega^2}{w_s t_s} \tag{3.18}$$

$$\frac{\partial \sigma}{\partial m} = \frac{g + A\omega^2}{w_s t_s} \tag{3.19}$$

$$\frac{\partial \sigma}{\partial f} = \frac{8\pi^2 mAf}{w_s t_s} \tag{3.20}$$

The applied displacement amplitude A is calculated before the experiment, using the relation $\sigma_a = mA\omega^2/(t_s w_s)$ with the desired stress amplitude σ_a (load amplitude from equation 3.13 divided by sample cross section). Therefore, it is interesting to calculate the error related to this calculation. For this purpose the following derivations are necessary:

$$\frac{\partial A}{\partial t_s} = \frac{\sigma_a w_s}{m\omega^2} \tag{3.21}$$

$$\frac{\partial A}{\partial m} = -\frac{\sigma_a w_s t_s}{\omega^2 m^2} \tag{3.22}$$

$$\frac{\partial A}{\partial f} = -\frac{\sigma_a w_s t_s}{2m\pi^2 f^3} \tag{3.23}$$

Provided the variables contained in the stress σ (i.e. in the calculated amplitude A) are stochastically independent, the Gaussian error propagation can be used to calculate the standard deviation STD_σ and STD_A out of the input parameters (see chapter 4).

3.2.2. FEM-simulations

Finite Element simulations of the system were conducted, in order to support and to enlarge the analytical investigations presented above. By means of modal analyses, it could be verified that no Eigenmode appears at a frequency close to the excited push-pull mode of the sample. In order to represent the continuum behavior of the components as accurate as possible, a model with very high mesh quality was established for the first studies. Later on, the final setup configuration was validated with a more detailed finite element model in terms of geometry, while having a less sophisticated mesh quality.

The volumes and boundary conditions of the first model are represented in figure 3.10. The effort of calculation is reduced using the symmetry of the problem by building a quarter model. The piezo actuator consists of a housing, a piezo stack and a spring, which preloads the piezo stack in compression. For this purpose a temperature is applied which expands the spring until the specified preload is reached, before starting the modal analysis. The geometry of the piezo actuator corresponds to the technical documentation of the manufacturer. The clamp is represented as a revolution part (intermediate development design), the sample is a sheet having rectangular shoulders at the transitions from the gage to the grip sections and the mass is a cube. The materials used for the different components are given in table 3.2 with their properties. All bodies were meshed with three-dimensional hexaeder elements (Ansys: solid45) with eight nodes, having three degrees of freedom each. (In order to guide the automatic sweep function of the program, two surfaces were meshed with two-dimensional four-node-elements (Ansys: plane42) during the modeling process.) The whole model consists of 51594 elements.

material	**Young's modulus [GPa]**	**density [kg dm^{-3}]**
steel	210	7.8
PZT	100	7.5
aluminum bronze	110	7.8

Table 3.2. Material properties used in the FE-model.

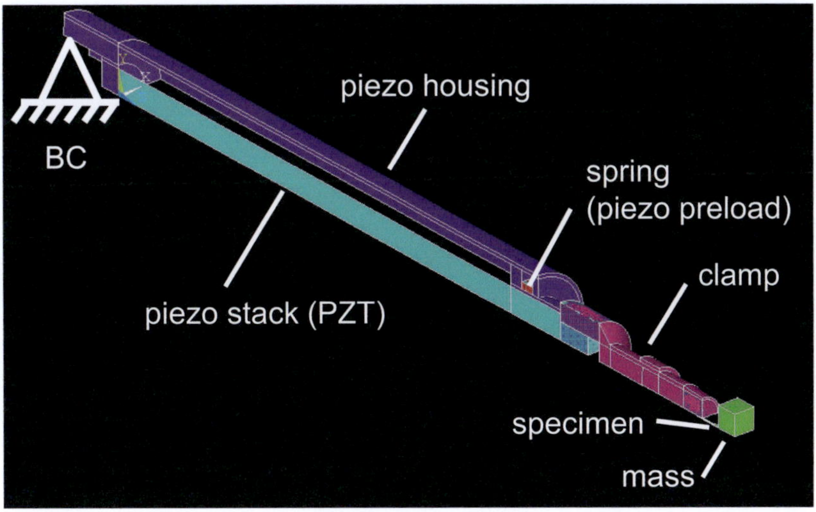

Figure 3.10. Quarter FE-model to study Eigenmodes as well as stress and displacement distributions in the piezo actuator and in the sample. Size scale: the piezo housing is 91 mm long.

Details of the mesh are shown in figure 3.11. Mainly hexagonal elements were created in order to reduce errors in calculation due to bad shaped elements. All conducted FE-analyses were purely mechanic. The piezoelectric behavior (multi-physics analysis) was not investigated.

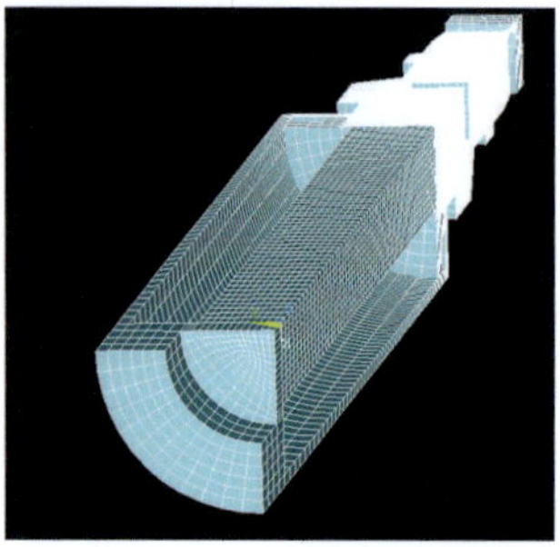

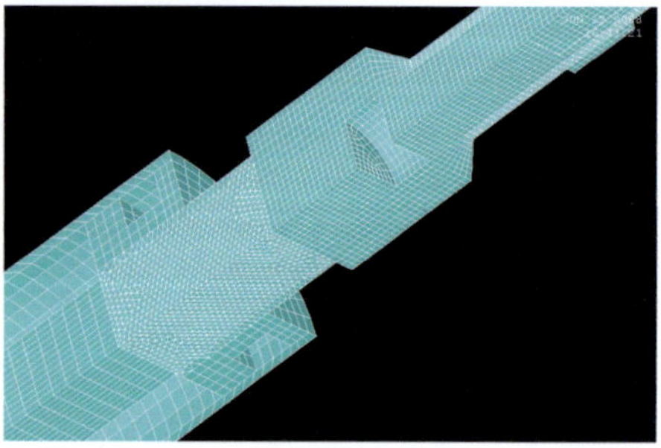

Figure 3.11. Details of the mesh: hexagonal elements were used as far as possible in order to get a high quality of the stress calculations. Size scale: the cross section of the quarter piezo ceramics stack is 5 mm × 5 mm in size. (Left) View along the piezo actuator in the direction of the mass (at the top of the picture); (Right) Transition between piezo actuator and clamp.

A modal analysis of the piezo actuator without the other setup components showed a natural frequency of the actuator's first mode at 8270 Hz. This is in the range of 8.5 kHz±20 % as specified by the manufacturer. Thus, the FE-model of the piezo actuator is considered to be valid.

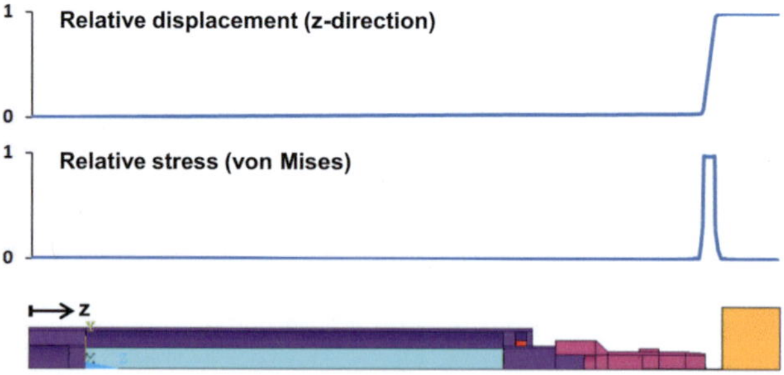

Figure 3.12. Stress and displacement are highest in the sample and negligible in the piezo actuator. Size scale: the mass (rectangle on the right) is 10 mm long.

The relative stress and displacement distribution from a modal analysis of the whole model (piezo actuator, clamp, sample and mass) is shown in figure 3.12. For this purpose, along the symmetry-axis of the model (z-axis) the relative stresses and displacements are plotted over the position z. Stress and displacement values are taken from the elements which have the smallest distance to the z-axis and divided by the maximum value found by this method. It can be seen that stresses and displacements are maximum in the sample and that the solicitation of other components can be neglected. Consequently, it is expected that sample and mass do not notably affect the mechanical behavior of the piezo actuator so that the system behaves very similar to the mass spring oscillator model with an infinite stiff piezo actuator presented above (see figure 3.9). This FE-model with all major setup components helped to find the amount of the mass needed to get a push-pull Eigenmode at about 1 kHz. Since the Eigenfrequencies found in experiments agree well with the frequency level predicted by the simulation, the FE-model with all setup components is validated.

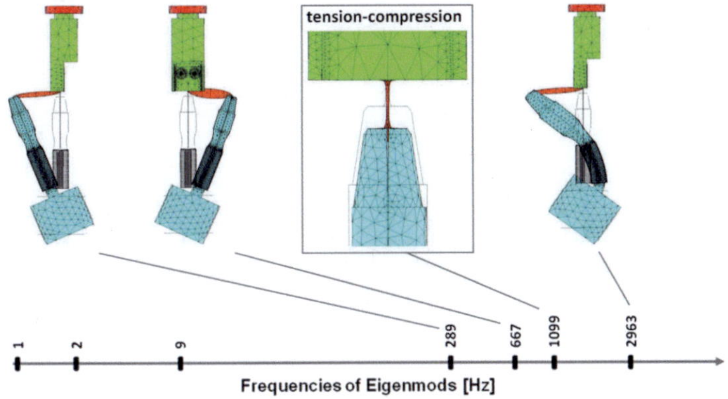

Figure 3.13. Eigenmodes with their natural frequencies: Tension-compression at 1099 Hz. The displacement is magnified for a better visualization. Size scale: the upper clamp (green) is 12 mm wide and 10 mm thick.

All the manipulations necessary to mount the sample and to prepare the experiments required a more complex design of the components. This design was investigated in further FE-analyses. A modal analysis showed that the frequencies for other Eigenmodes of the system are apart from the push-pull mode at which the system is excited during the experiment (see figure 3.13). Thus, it is expected that the sample is mainly loaded in push-pull direction when excited at this frequency.

3.2.3. Final design

The basic idea of the resonant setup was presented in figure 3.8 showing the first setup design. In order to facilitate the mounting procedure of the sample, to reduce the space taken by the setup (one optical table) and to reduce the noise emitted by the vibrations a new design was developed (see figures 3.14 and 3.15).

Figure 3.14. Final design of the resonant setup: (Left) Computer and soundproof box containing the setup; (Right) insight into the soundproof box showing the resonant setup.

The setup is mounted onto a bread board fixed to the wall. The bread board is supported by rubber elements in order to isolate it mechanically from the environment. A noise canceling box encloses the setup.

The piezo actuator is fixed to a cantilever, which is optimized not to vibrate at the working frequency of the setup. Between most of the components a layer of cork is added in order to reduce the transmission of solid-borne sound. A special support, which carries the displacement sensor and provides a clamp for mounting the mass was designed. It sits on an x-y-z-rotation stage and has an accurate screw necessary to adjust the height of the displacement sensor. A camera (Pixelink, Canada) with an optical adapter (infinitube) for Nikon microscope objectives is mounted onto an x-y-z stage. Therefore the sample surface can be observed during experiments in order to detect damage. An overall view of the sample with mass is given during the experiments by means of a webcam and a 3 W-light source inside the noise canceling box.

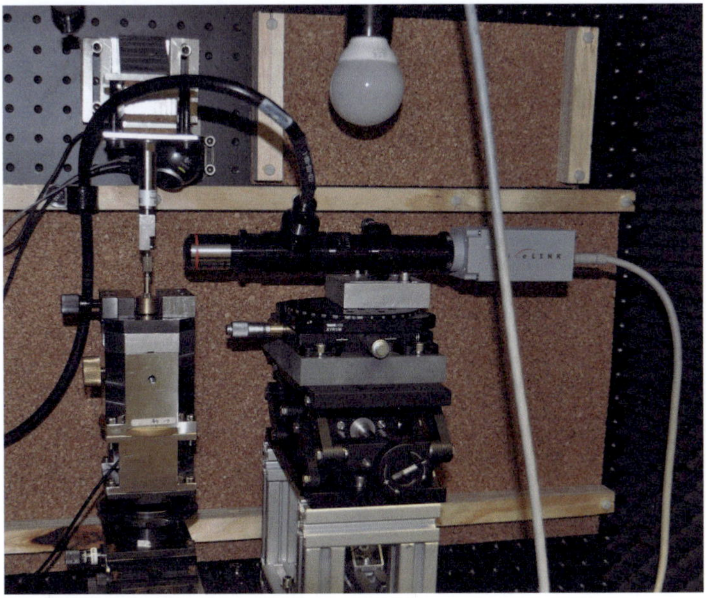

Figure 3.15. Final design of the resonant fatigue setup: mass positioning system with integrated displacement sensor, camera on its separate positioning system, piezo actuator on optimized cantilever.

3.2.4. Experimental procedure

At first, the sample is mounted onto the clamp (optically aligned with magnifying glass), which is fixed to the piezo afterwards. In the next step, the piezo carrying the sample is screwed onto the setup. During this procedure, it is rotated until the sample surface is perpendicular to the optical axes of the camera. The mass is clamped onto the x-y-z-rotation stage, which contains the displacement sensor, and is then positioned beneath the sample. Afterwards, the mass is lifted in the z-direction, until its gripping system surrounds the clamping surface of the sample. This gripping system is then closed manually. During this procedure, the position of the gripping system with respect to the sample is optically checked, using the camera and a magnifying glass. If necessary, it is corrected with the x-y-z-rotation stage in order to prevent undesirable deformation of the sample. When the gripping system is closed, the mass is released from the x-y-z-rotation stage, i.e. freely suspended. At this stage, a frequency sweep is conducted with the piezo actuator at low amplitude. The response of the mass is measured in order to detect the frequency of maximum amplification (response amplitude of the mass divided by the excitation amplitude of the piezo). This frequency is used for the fatigue experiment. The displacement amplitude of the mass necessary to obtain the desired load at this frequency is calculated using the equations introduced above. It is maintained at a constant level during the experiment. At the end of the experiment, the fatigue crack grows so fast that the controlled response amplitude of the mass decreases, despite the control system (loss in stiffness). After a few number of cycles, the amplitude decreases by 3 %, which has been found to be a good criterion to stop the experiment just before final failure. With this method, samples containing fatigue cracks at the end of their lifetime can be investigated with a microscope.

3.3. Data acquisition and control

Since the same hardware and software components are used for different setup configurations, the software is modular and sometimes kept quite abstract. Values, written into files are always recorded in volts and variables usually have a general name, such as AI0 for the value measured at the analog input AI0. For instance, this value can correspond to the voltage measured by the load cell (setup up to 100 Hz) or to the voltage measured by the displacement sensor (resonant setup).

All recorded files are post-processed in Matlab in order to create diagrams and to connect them to the image correlation results (calculated in Matlab), in case optical strain measurements are performed. Conversion of the voltage values into units of the measured parameters is done during the post-processing in Matlab. An important advantage of Matlab is the amount of data which can be analyzed and visualized. It is possible to draw graphs of arrays containing several millions of data points which is advantageous for a detailed analysis of long time fatigue experiments.

3.3.1. Measuring concept - Hardware

The interaction between hardware components of the resonant setup is shown in figure 3.16. They are all connected to a personal computer (PC) having an integrated Field Programmable Gate Array (FPGA).

The FPGA-board realizes fast and stable data acquisition, signal generation as well as partial data treatment. It is a PCI board mounted into a standard quad core PC (so called host computer). Having its own processor on board, which serves as a precise timer, the FPGA works at a frequency of 40 MHz. Connections between the processor and logical components such as flip flops,

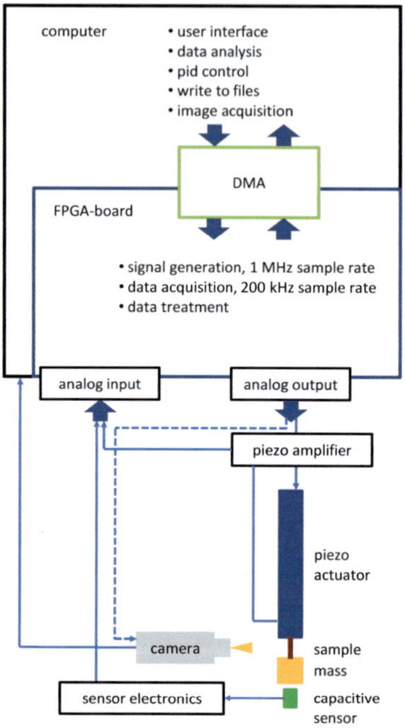

Figure 3.16. Schematic of the interaction between hardware elements. The setup components are driven by a personal computer with an integrated FPGA-board, which provides a resolution of 16 bit per 20 V for all analog input and analog output channels.

look up tables and Boolean gates can be changed by programming. Thus, the FPGA works similarly to a hard wired board specially designed to fulfill measurement and control of the fatigue setup, independently of the host computer. The FPGA provides a high level of parallelization, timing accuracy, speed (close to real time) and reliability. Eight analog outputs (AO0 to AO7, 1 MHz maximum sample rate), eight analog inputs (AI0 to AI7, 200 kHz or 750 kHz maximum sample rate depending on the used model) as well as numerous digital ports work in parallel and enable precise measurement and control during the fatigue experiments. The available range for all AIs and AOs is −10 V to 10 V. Transduction between digital values and AI/AO values on the FPGA is conducted with a resolution of 16 bit (corresponding to about 0.305 mV per digit). Therefore, all analog voltage signals are represented by 16 bit integer values in the software program running on the FPGA (see below).

Data is transferred between FPGA and host computer via direct memory access (DMA). User interface, PID-control as well as analysis and record of the data are implemented on the host computer with LabView (see below). Images are acquired by use of the camera software which runs also on the host computer.

The signal generated at the analog output AO0 is connected to the piezo amplifier (Physikinstrumente, Germany) and induces the motion of the piezo actuator. A position sensor integrated in the piezo actuator returns a signal to the amplifier, which converts it to a voltage in the range of 0 V to 10 V (corresponding to the travel range of the piezo). This voltage is measured at the analog input AI1 and indicates the current piezo position (6 V µm^{-1} and 3 V µm^{-1} for the piezo actuators with 60 µm and 30 µm travel range respectively). The amplifier provides an integrated control of the piezo position, in order to compensate the nonlinear behavior of the piezo actuator. For experiments in the kHz-regime where this control is too slow and for load controlled experiments (load control is realized by an implemented PID-control on the computer) where instability can appear, the position control of the piezo amplifier is switched off.

The capacitive sensor (200 µm range of measurement) works with a high frequency alternating voltage. It is connected to an electronics module (called sensor amplifier in figure 3.16) which can be set to a maximum output frequency of 50 kHz. For this frequency, the specified resolution of the sensor is 0.2 µm. The sensor amplifier is connected to AI0 of the FPGA-board and generates $0.05\,\mathrm{V\,\mu m^{-1}}$ distance to the mass fixed at the bottom of the sample.

The camera is connected to the fire-wire port of the computer. In the experiments presented in this work, it was triggered by the camera software. For a very accurate trigger signal driven by the FPGA-board, the camera can also be connected to an output of the FPGA as indicated by the dashed line. For this purpose, software was developed.

For the setup working up to 100 Hz figure 3.16 is the same despite the capacitive sensor (and its electronics) is replaced by the load cell (and its electronics), and in addition to the camera the motor is connected to the PC by means of a PCI-board.

3.3.2. Software written with Labview

The software, which drives the custom built setups, was written in the graphical computer language LabView. Sections of the program, which receive and transmit information from and to other sections, are called Virtual Instruments (VIs, comparable to functions or methods in other languages). They are grouped in a project, where also the hardware (FPGA) is integrated and can be configured. In figure 3.17 (p. 97), the five main VIs of the project driving the setup are illustrated.

The FPGA-VI contains the software section which is running autonomously on the FPGA-board. It has to be compiled in a time consuming process after each software change during which the VI is translated into the logical elements available on the FPGA-board. The Host-VI is the most important program on the host computer. It starts the FPGA and acts as a convenient user interface. It should be noticed that for hardware reasons, all variables in the FPGA-VI are integers. For instance, the voltage values between -10 V and 10 V measured at the analog inputs are returned in 16 bit values and the loop frequencies are given in integers in terms of multiples of the FPGA-clock frequency (40 MHz). These values are converted into standard units in the Host-VI.

The Motor-VI controls the motor via the serial port of the motor controller board and the Write-Analog-Input-Data-VI writes the data collected by the FPGA from the analog inputs into an ASCI-file. Both VIs can be switched on and off without affecting the HOST-VI. Variables which can be accessed from all independently running VIs in a project are called global variables (in distinction to local variables only accessible by the owning VI) and are collected in the Global-Variables-VI.

In figure 3.17, the communication flow between the five in parallel running VIs is indicated by arrows. Each VI consists of continuously parallel running loops (called F1 to F3 for the FPGA-VI, H1 to H3 for the Host-VI etc.). The aim for having parallel loops is usually accelerating the software, sparing computer resources (repeat loops only as often as necessary) or triggering processes which must run at a certain frequency. The frequency at which each main loop runs is shown in the upper right corner of the rectangle representing the loop. Numbers indicate a frequency in Hz, letters represent variables which will be explained in the following.

Data acquisition on the FPGA-board

Loop $F1$ runs at the frequency f_m, which is also measurement frequency. All eight analog inputs (AI0 to AI7) and all digital inputs are read out in parallel at this rate and the values are written together with a sample number into a DMA-buffer, which can be accessed by the Host-VI. Since the frequency f_m is known, the sample number is equivalent to a time stamp (time zero when the measurement starts). The user can choose f_m in the Host-VI in Hz (e.g. in order to limit the data which must be handled by the host computer to the necessary amount). In addition, oversampling is available for f_m lower than 100 kHz, which is half the maximum sample rate of the FPGA. A number of data points can be chosen, which are acquired at the maximum sample rate and averaged on the FPGA-board. For this purpose a special routine was developed, since calculations must be done in the integer format and variable overflows must be prevented. The routine is explained in the following.

The voltage signal $y(t)$ connected to an analog input varies over the time t and is measured at the maximum sample rate. For the averaging process the measured 16 bit values y_i acquired at the time t_i must be summed up and divided by their number N.

$$\bar{y} = \frac{1}{N} \sum_{i=1}^{N} y_i \tag{3.24}$$

Herein, $\bar{y}$ is the mean value of N data points acquired by oversampling. The division by N can be performed on the host computer. Summing up the data on the FPGA-board, a variable overflow must be prevented. Therefore, the first measured value is taken as a reference and only the differences between this reference and every following data point are summed up. This leads to a smaller value which must be transferred to the host computer. For this purpose equation 3.24 can be transformed into the following form:

$$\bar{y} = y_1 + \frac{1}{N}\sum_{i=2}^{N}[y_i - y_1] \tag{3.25}$$

For each data point $\bar{y}$ three values are transferred to the host computer: The first measured value y_1 without averaging, the average of the differences between the first value and every following value (sigma sign in equation 3.25), and an overflow check. The latter is usually not active because the measured value changes barely during a period of oversampling. However, each time when the sum of all $(y_i - y_1)$-terms exceeds the range of the 16 bit integer variable in the positive direction, the overflow check is increased by 1. In the negative direction it is decreased by 1. Hence, the host computer can take every overflow into account.

In addition to the data acquisition a simple data analysis is conducted in loop $F1$. The acquired value of analog input AI0 which corresponds to the load applied on the sample during the experiments is compared to an admissible range defined by a maximum and a minimum value. When the signal drops out of this range, a variable called *emergency_stop* is set to true which stops the motion of all actors in the setup.

Loop $F1$ can only run at 200 kHz, when data acquisition and data transfer to the DMA-buffer are executed in parallel. For the experiments presented in this work, a sample rate of 50 kHz was sufficient. Therefore, a non parallelized FPGA-VI, which does not work up to 200 kHz, was used.

Signal generation on the FPGA-board

Loop $F2$ generates the output signal at analog output AO0. This output signal induces the motion of the piezo actuator of the setup. The piezo actuator with 60 µm travel range moves with the ratio $r_p = 6\,\mu m\,V^{-1}$ generated at AI0. The loop runs at the frequency f_v which depends on the velocity v at which the piezo actuator should move linearly. When the user changes the velocity in the Host-VI, the frequency f_v is calculated at which the 16 bit output must be incremented by 1 digit (smallest output change in voltage that can be realized by the FPGA-board $\Delta y_{min}^{AO} = \left(20/2^{16}\right) V)$ in order to obtain the desired velocity:

$$f_v = \frac{v}{r_p}\frac{2^{16}}{20} \tag{3.26}$$

Setting f_v in equation 3.26 to the hardware limited maximum output frequency of 1 MHz and solving for v delivers the maximum velocity $v_{max,smooth}$ at which the piezo actuator with 60 µm travel range can be driven as smooth as possible, that is to say with steps of the smallest possible output increment Δy_{min}^{AO}. This increment corresponds to a piezo motion of 1.83 nm. Thus $v_{max,smooth}$ is found to be 1831 µm s^{-1}.

In each loop iteration of loop $F2$ the current output value y^{AO0} is subtracted from the target output value. According to the sign of the result, the output value is incremented or decreased by 1 digit. When the difference is zero, the target is reached and the output is not modified. When the piezo actuator is driven in the load controlled mode, the same mechanism works, unless the current input value y^{AI0} is subtracted from the load target. It should be noted that the piezo drives always with the chosen velocity, also in the load controlled mode. When the load control becomes instable the piezo velocity should be lowered. As long as the variable *emergency_stop* is true or the variable *stop_control* is zero, the output value y^{AO0} is not modified even though it differs from the target value (or y^{AI0} differs from the load target in the case of the load controlled mode).

Inside loop $F2$, there is loop $F3$ for the generation of sinusoidal signals, i.e. sinusoidal piezo motions. When the variable *wave* is set to true, loop $F2$ receives the mean value of the wave (called offset) as a target. As soon as the target is reached, the program enters loop $F3$, which generates the sinusoidal wave until either *wave* is set to false or the *emergency_stop* is set to true. Loop $F3$ always runs at the maximum output frequency of 1 MHz.

Host-VI - Data transfer to the Host Computer

Starting the Host-VI, the user chooses at first the names and locations for the files stored by the Host-VI. Thereafter, the sample rate f_m and the number of data points used for oversampling $(N-1)$ are defined. At the moment where the *start_ daq*-button is pressed, these parameters are transmitted to the FPGA-VI and the measurement is started. Furthermore, the current system time of the host computer is written into a global variable, so that the start time of the measurement can be stored in a file. Thus, the sample numbers created by the FPGA (time stamp relative to the start of the data acquisition) can be converted to the system time of the computer.

Loop $H1$ in the Host-VI is executed twice per second. In each iteration the data contained in the DMA-buffer (data block) is read out, converted into standard units and passed to a global variable. Afterwards, the data block is analyzed. The mean value, the frequency and the amplitude of AI0 and AI1 as well as the phase angle between both signals are calculated. Some of these values are stored with the time stamp of the first sample of the data block in an ASCI-file (see *fr.dat* in table 3.3). The mean value and the amplitude of AI0 are passed to a PID-control-VI when the load control is active during wave generation. Thus, the mean value and the amplitude of the sinusoidal load applied on the sample can be adjusted two times per second. If an adjustment is necessary, the new amplitude and offset values returned by the PID-algorithm are written into a file (see *ctr.dat* in table 3.3).

file name	sample rate	recorded data	VI
fr.dat	2 Hz	sample number, frequency(y^{AI0}), offset(y^{AI0}) amplitude(y^{AI0}), phase angle(y^{AI0}, y^{AI1})	Host
ctr.dat	≤ 2 Hz	sample number, offset(y^{AO0}), amplitude(y^{AO0})	Host
cyc.dat	OD	sample number, 2× cycle number	Host
load.dat	$\leq f_m$	sample number, y^{AI0}, y^{AI1}	Write
log.dat	1 Hz	time stamp, motor position	Motor

Table 3.3. The data is written into files (file names are chosen by the user) at different sample rates (OD: on demand, f_m: frequency of measurement). The table shows the data contained in each file as well as the VI which creates the file (Write: Write-Analog-Input-Data).

In loop $H2$ current data (only one sample per DI and per AI) is read from the FPGA as fast as possible (maximum frequency that can be realized by the Host-computer). The frequency of this loop is arbitrary and not every data point acquired by the FPGA is treated in this loop. However, value changes can be recognized very fast by the HOST-VI in this loop. This is used for a real time visualization of the data on the screen and for a fast execution of automatic data dependent software behavior, as will be explained later.

User interface control in the Host-VI

The user interface is handled in loop $H3$. It is executed on demand, i.e. when the user presses a button or changes some values in the user interface. Mainly, the motion of the piezo actuator is started and stopped, which is related to several software actions. Parameters defining the motion (e.g. speed and target position) must be passed to the FPGA-board and the number of generated cycles are written into a file (see *cyc.dat* in table 3.3) as soon as a sinusoidal wave is stopped. In general there are three different kinds of motion, the user can ask for: A linear motion at a defined speed (see explanations for loop $F2$), a sinusoidal motion at a defined frequency with the desired values for amplitude and mean value (called offset, see loop $F3$), and a slow wave, which is equivalent to an automatically driven linear to an fro motion (triangular or trapezoidal wave, defined by the speed of the linear motion and by the dwell time). For frequency and amplitude, a sweep at a desired ramp speed is available. Linear and sinusoidal motion can both be executed with or without load control.

When a load controlled wave is started, the piezo actuator moves at first linearly to the mean value of the sine and starts an uncontrolled wave with the desired frequency at an amplitude specified by a variable *start_wave_amp* (usually chosen to be very small). Thus, the VI measuring the wave parameters (frequency, amplitude and offset) can return values which are necessary as an input for the PID-control. Afterwards, the load control which adjusts amplitude and offset can be started. This procedure is a good example for an automatic program behavior depending on the measured values, which is handled in loop $H2$. Further examples are stopping a cyclic motion after a desired number of cycles, changing the target values to induce the to and fro motion for a slow wave, when the upper or lower target value is reached, and handling the emergency stop on the host computer. When the variable *emergency_stop* changes on the FPGA side, this information is distributed by the Host-VI, so that all components of the setup which are connected to computer ports (also those which are not connected to FPGA ports), such as the motor (at the serial port), are stopped as fast as possible.

Write analog input data into a file

The Write-Analog-Input-Data-VI contains only one loop $W1$ which writes the raw data acquired at the two analog inputs AI0 to AI1 together with the FPGA-time-stamp into an ASCI-file (see *load.dat* in table 3.3). The loop is executed at a frequency of 2 Hz, which corresponds to the rate at which a data block is read out from the DMA-buffer by the Host-VI. In summary, the raw data is acquired by the FPGA, read out block wise from the DMA-buffer by the Host-VI and written into a global variable after conversion into standard units. Thus, the data can be accessed from every VI in the project. Several versions of the Write-Analog-Input-Data-VI exist which can be switched on and off according to the needs of the user. One version writes all the acquired data into a file, another version enables the selection of a write frequency which differs from the frequency of measurement, a third version records only a defined part of each data block and a fourth version writes a defined part of the data block every j seconds, where j can be chosen by the user. For VHCF-experiments j is usually set to 1800 s. These versions were developed to reduce the amount of data which is written to the hard disk, in order to reduce the need of memory and to facilitate the post processing.

Motor control in the Motor-VI

The Motor-VI consists mainly of three loops, as can be seen in figure 3.17. At a frequency of 2 Hz the history diagram showing load over total piezo plus motor displacement is updated in loop $M1$. User interactions concerning the motor command (e.g. change of speed or change of target position) are handled in loop $M2$. When required by the user, a loop $M3$ is running which writes the motor position over time into a file (see *log.dat* in table 3.3).

This function is of special interest for experiments with samples showing a large deformation. In this case, the travel range of the piezo actuator is not sufficient and the motor must be moved during the experiment. In a further loop which is not represented in the figure, the motor position is polled at a rate of 500 Hz.

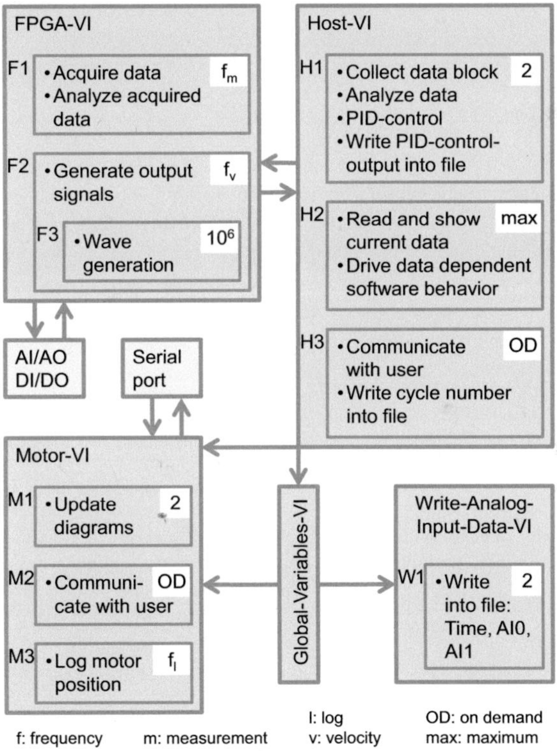

Figure 3.17. Schematic of the communication between five software parts (VIs) as well as analog, digital and serial ports. Frames represented inside the VIs are named F1, F2, ... in the FPGA-VI; H1, H2, H3, ... in the Host-VI; etc. and show permanently running loops. The distinct loop frequency is indicated in the upper right corner in terms of a variable or in terms of a number in Hz. Arrows indicate the flow direction of information. The FPGA-VI runs on the FPGA-board, the other VIs run on the host computer.

4. Mechanical tests on Micro Molded Aluminum Bronze

4.1. Experimental - Micro Molded Aluminum Bronze

This chapter is focused on the investigation of the high cycle and very high cycle fatigue regime as well as on the damage morphology caused by fatigue of micro molded aluminum bronze. For this purpose, high frequency experiments in the kHz-regime were performed using the resonance setup presented in chapter 3. At higher amplitudes, fatigue experiments were also conducted at 25 Hz ($R = 0.1$) down to the low cycle regime. These results can be compared to the work of Rögner [55], who fatigued samples (also at 25 Hz, $R = 0.1$), which were produced through the same process route at a lower mold temperature.

Tensile tests on samples manufactured with the same mold temperature (namely 1000 °C) as the samples from the present study are presented in [55]. Thus, only few quasistatic tests were conducted to confirm the results from the literature.

The mechanical behavior of the different phases was investigated by nanoindentation. Optical microscopy, confocal laser microscopy, surface electron microscopy (SEM) combined with a focused ion beam (FIB) and transmission electron microscopy (TEM) were used to investigate the surface quality, the morphology of the phases in the material as well as their interaction with damage formation during fatigue. Since very small (nm-regime) and complex phases are present, several special techniques such as scanning transmission electron microscopy (STEM) combined with energy dispersive x-ray spectroscopy (EDX) and energy filtered TEM were applied with the kind help of the Laboratory for Electron Microscopy (LEM) at the Karlsruhe Institute of Technology (KIT). Also EDX in a SEM was used to investigate the distribution of elements and phases in the samples.

4.1.1. Investigated samples

Micro tensile samples were molded using investment casting of a commercial aluminum bronze alloy (CuAl10Ni5Fe4 named 2.0966; The tolerances for the composition are specified in the German standard DIN17665 in weight % as follows: Al 8.5-11.0, Ni 4.0-6.0, Fe 2.0-5.0, Mn 0.0-1.5, Zn 0.0-0.5, Cu balance). The cross section of the gage is 260×130 µm^2. All sample dimensions are shown in the drawing of figure 4.1 (Left). An as-cast sample observed through an optical microscope is presented in the center of the same figure. A detailed view of the gage section can be seen next to it in an image acquired by means of a confocal laser microscope (Keyence VK-9710k). The gage section of a polished sample (acquired with the confocal laser microscope) is shown in figure 4.1 (Right). An important manufacturing parameter is the mold temperature which was 1000 °C for all samples tested in this work. One batch consists of 18 samples arranged in a line and molded in one part. Before testing, each sample was inspected on all surfaces through an optical microscope with camera (1.75× to 14× magnifying lens system corresponding to 0.5 and 20 pixels per µm respectively), in order to check its quality. Usually, some samples of a batch have large defects at the surface or are bent due to manufacturing. Such samples were not used for the experiments. The width and the thickness of the samples were measured from images taken with the camera on the optical microscope (14× and 35× magnifying lens system corresponding to 4 and 10 pixels per µm respectively). Thus, the cross section could be calculated to determine the stress from load measurements. The gage roughness of the as-cast sample presented in figure 4.1 (Center) was determined with the confocal laser microscope and is $R_t = 2.64\,\mu m^*$ and $R_z = 1.49\,\mu m^\dagger$. This is in the range of

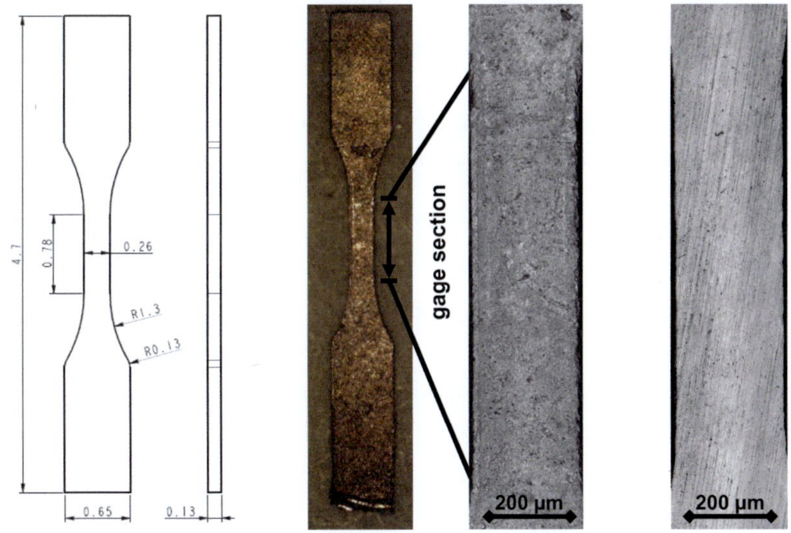

Figure 4.1. Tensile samples made of aluminum bronze: (From left to right) Drawing with dimensions shown in front and profile view; as-cast sample in the optical microscope and its gage section observed in the confocal laser microscope; polished gage section observed in the confocal laser microscope.

$^*R_t$ according to the standard ISO 4287
$^\dagger R_z$ according to the standard JIS B0601 1994

Most of the samples were tested as-cast in order to characterize the mechanical behavior comparable to micro parts, without modifying the surface roughness after the molding process. Also, polished samples were examined, in order to reduce the risk of surface roughness induced crack initiation during fatigue. Furthermore, small extrusions and intrusions which typically appear during very high cycle fatigue can be discovered more easily on a polished surface. Polishing of the front and the back side was done mechanically with diamond paper (down to a grain size of 1 µm). The side surfaces stayed as-cast, because mechanical polishing is not possible in this region and no well established process for electro polishing was found. Aluminum bronze is known to be very corrosion resistant and etching usually does not attack the different phases homogeneously. Thus, the number of notch effects due to the surface roughness, which can degrade the resistance to fatigue, is reduced, since they are more or less eliminated on the polished sample surfaces. However, the sizes of notches which still exist on the side surfaces are expected to be of the same order as the ones removed by polishing.

4.1.2. Quasistatic tests

Tensile tests were conducted on as-cast samples with setup 2 shown in figure 3.2. The samples were mounted as described in section 3.1.2 for the conventional flat sheet dog bone geometry. Strain was measured with digital image correlation (see section 3.1.1).

Nanoindentation was conducted on etch-polished samples with a nanoindenter XP (MTS Systems Corp., Eden Prairie, MN, USA) using a Berkovich tip. The indentation was performed at a constant strain rate (controlled with continuous stiffness measurement (CSM)) of $0.05\,s^{-1}$ to a depth of

1 µm. The indentation sites were chosen in a way to hit areas of different phases visible through the camera of the nanoindenter in order to distinguish their mechanical behavior. The following three types of regions were chosen for indentation: light regions showing almost no precipitates on the polished surface (matrix), dark etching regions and regions showing globular or dendritic-shaped precipitates. The hardness and the Young's moduli were determined from the load displacement information gained during indentation.

4.1.3. Cyclic tests

Setup 1 was used for fatigue tests at frequencies up to 100 Hz (setup with load cell, see figure 3.1). These tests were load controlled and mainly conducted at 25 Hz (for three samples also 2.5 Hz and 100 Hz were applied). For some samples, the first few cycles were done at quasistatic conditions (strain rate in the range of $1 \times 10^{-4}\,s^{-1}$) in order to measure strain with the image correlation. A tension-tension load with the load ratio $R = 0.1$ was applied.

Higher frequencies were applied to reach the very high cycle regime in a reasonable amount of time. Fatigue tests up to 2 kHz were performed using the resonant fatigue setup explained in section 3.2. The system is excited at the natural frequency for the tension-compression mode of the sample. This frequency is adjusted by the mass fixed to the sample. Thus the R-ratio is close to -1. For experiments at about 1 kHz, masses of about 30 g were used (30 g, 32.6 g and 36.6 g). For experiments at about 2 kHz, a mass of 7.2 g was applied. Since the low frequency pendulum mode (sample bending) was excited for some samples in this configuration (no fatigue test could be performed in these cases), only few tests were conducted at the 2 kHz frequency regime. A mass of 21.2 g was used to adjust the experiments on polished samples to the kHz-regime. These samples have a reduced stiffness in tension-compression, because their cross section is reduced. Consequently, the kHz-regime is obtained with a lighter mass compared to as-cast samples.

The error in stress measurement and in determination of the applied displacement amplitude was estimated using equations 3.17 to 3.22 for a typical experiment as shown in table 4.1.

variable x_i	**mean** $\bar{x}_i$		**std** STD_{x_i}		$\partial\sigma/\partial x_i$		$(\partial\sigma/\partial x_i)STD_{x_i}$	
thickness t_s	130	µm	2	µm	−1.36	MPa µm^{-1}	−2.7	MPa
width w_s	260	µm	2	µm	−0.68	MPa µm^{-1}	−1.4	MPa
amplitude A	4.5	µm	0.04	µm	37.38	MPa µm^{-1}	1.5	MPa
mass m	32	g	0.1	g	5.55	MPa g^{-1}	0.55	MPa
frequency f	1000	Hz	0.1	Hz	0.34	MPa Hz^{-1}	0.03	MPa

Table 4.1. Errors in stress measurement induced by the respective variable x_i are calculated using the derivative of the stress σ with respect to x_i and the standard deviation STD_{x_i}.

Deriving the displacement amplitude with respect to the variables gives $(\partial A/\partial x_i)STD_{x_i}$-values which are one order of magnitude smaller than the corresponding $(\partial\sigma/\partial x_i)STD_{x_i}$-values. Using the Gaussian error propagation (see [87] (p. 788)), the resulting standard deviation of the measured stress applied on the sample is $STD_\sigma = 3.2\,\mathrm{MPa}$ (Euclidean norm of the elements contained in the last column of table 4.1).

4.1.4. Analysis of microstructure and damage

The overall microstructure of the samples was investigated in detail by Rögner, Baumeister and Okolo (see section 2.4.2). The focus of the present work was to investigate the local microstructure of cracked and fractured samples in order to reveal the interaction between the local microstructure, the damage, and fracture behavior. The main work was conducted at a dual beam work station (FEI, Nova Nanolab 200 surface electron microscope (SEM) and focused gallium ion beam (FIB)). The fracture surfaces of samples fatigued until rupture were observed. Furthermore, samples containing fatigue cracks but not being separated into two parts were investigated. For this purpose, the microstructure close to the cracks was revealed by means of focused ion beam cuts. A platinum layer of about 3 µm thickness was deposited at the site of interest, in order to obtain a smooth cut. Material was removed with the focused ion beam. With this method the underlying microstructure is revealed.

In order to assign the different phases visible with the SEM and with the FIB to the phases reported in the literature, EDX was conducted in an SEM (Zeiss, LEO 1530 with Thermo-Fisher NSS-analyzer). With this technique, very fine phases could not be resolved. Therefore, further analyses including dark field imaging, bright field imaging and EDX using scanning transmission electron microscopy (STEM) were conducted at the TEM (Philips, CM200 FEG/ST, 200 kV). In addition, electron energy loss spectroscopy (EELS) was performed at the TEM (Zeiss, LEO 922 Omega, 200 kV), in order to investigate the distribution of the elements among the different phases. Finally, a TEM lamella containing a fatigue crack was prepared and observed in the TEM (FEI, Titan 80-300 300 kV). In addition to bright field and dark field imaging, STEM based EDX was performed. TEM lamellae were prepared at the FIB. A platinum stripe (about 15×1 µm) of about 3 µm height was deposited at the location where the lamella was extracted. The material around was removed by FIB cutting. Before separating the lamella entirely from the surrounding material, it was attached to a micromanipulator by means of platinum deposition. Thus, the lamella could be transferred to a lift-out-grid and attached again by means of platinum deposition. The lamella was thinned until electron transparency by FIB milling. The lift-out-grid was introduced into the TEM in order to observe the lamella.

4.2. Results - Micro Molded Aluminum Bronze

The results presented in this chapter shall help to understand the failure mechanisms of micro molded aluminum bronze, mainly during high cycle and very high cycle fatigue. Since this material contains several phases of different shapes, chemical compositions and crystallographic structures, they are likely to have a strong impact on the failure mechanism. Their mechanical behavior was investigated with nanoindentation and compared to the overall behavior of the material in quasi static tensile tests as shown in section 4.2.1. The fatigue behavior in the low cycle, high cycle, and very high cycle regime is shown in the following section. Before presenting the interaction between the microstructure and the fatigue damage in the last section, the microstructure itself is analyzed in section 4.2.3.

4.2.1. Monotonic loading - Hardness and tensile characteristics

Indents were made in regions of three different types, which can be distinguished in the optical microscope as follows: light regions showing almost no precipitates on the polished surface (matrix), dark etching regions and regions showing globular or dendritic-shaped precipitates of about 4 µm in diameter. SEM images acquired after the indentation experiments are represented in figure 4.2. They show indents into all three types of regions, which can now also be described as regions without precipitates (matrix), with lamellar precipitates (formally dark etching) and with globular (or dendritic) precipitates. It can be seen that the formation of the pile-up and of slip bands around the indents are different at locations containing different phases.

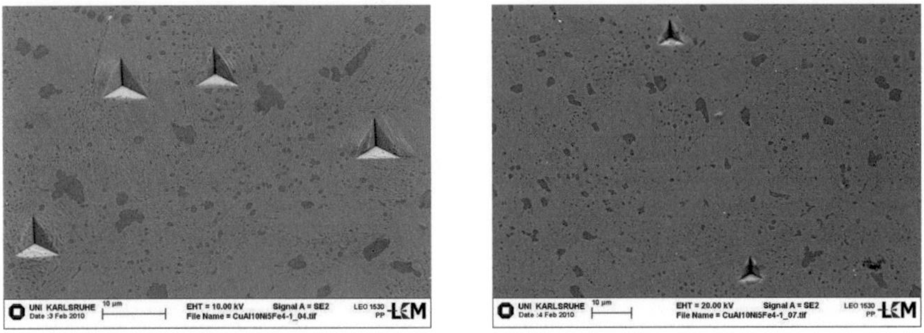

Figure 4.2. Nanoindentation at locations containing different phases (images acquired from LEM at KIT): (Left) Indents into regions without precipitates (matrix) and into regions with lamellar precipitates; (Right) Indents into globular precipitates.

The Young's moduli found from the unload curves are shown in figure 4.3 (Left) for each indentation. They vary between 100 and 120 MPa with an average of 112 MPa.

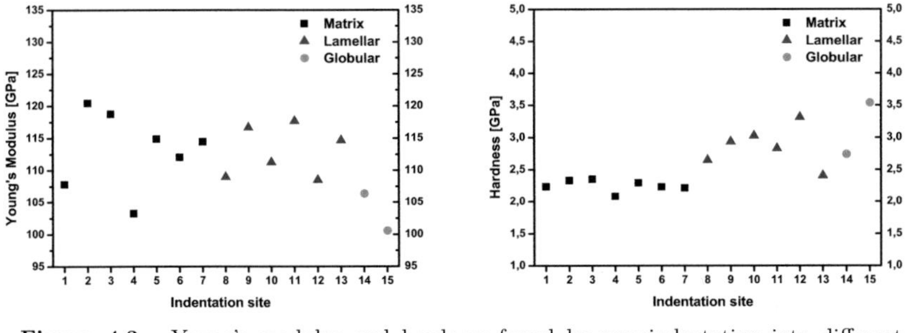

Figure 4.3. Young's modulus and hardness found by nanoindentation into different phases. The values were determined from the unloading segment of the indentation load displacement curve.

The hardness is presented in the same figure (Right). The matrix with 2.3 GPa tends to be softer than the regions containing lamellar (2.9 GPa) and globular or dendritic-shaped precipitates (3.1 GPa). The last-mentioned hardness is expected to be underestimated, because of a "sink-in" of the precipitates into the softer matrix, which was indicated by the load-displacement curve. For the same reason, the values of the Young's modulus measured from the indents 14 – 15 must be interpreted carefully.

Two stress-strain curves from quasi static tensile tests (strain rate about $5 \times 10^{-4}\,\mathrm{s}^{-1}$) are shown in figure 4.4. Both samples came from the same batch (cast piece). The yield strengths $R_{p0.2}$ are 565 MPa and 606 MPa and the ultimate strengths are 808 MPa and 848 MPa respectively. Young's moduli of 107 GPa and 98 GPa were determined from the stress-strain curves. The fracture strains of 8 % and 9 % are a measure for the overall ductility of the material.

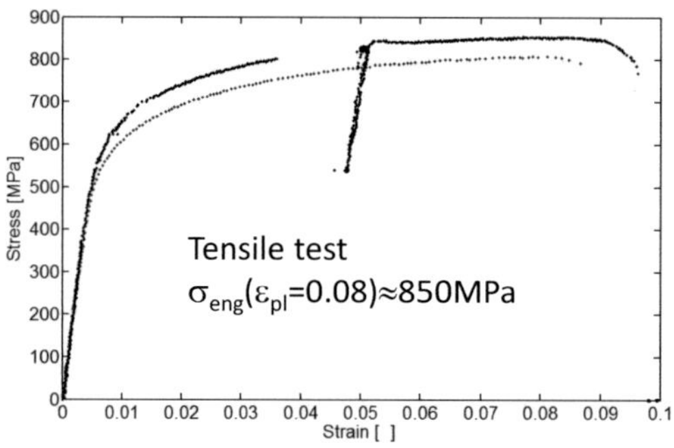

Figure 4.4. Examples for the tensile characteristics of the micro molded Aluminum bronze.

4.2.2. Fatigue behavior

A lifetime diagram containing the results of all fatigue tests is shown in figure 4.5. The investigated samples come from seven different batches produced with the same parameters. The triangular shaped data points are gained with setup 1 (see figure 3.1) at a load ratio of $R = 0.1$. As expected, increasing lifetimes for decreasing load amplitudes are observed in the low cycle regime. Most of these experiments were conducted at 25 Hz. Single tests at 100 Hz and 2.5 Hz shown in the same diagram can help to estimate, whether the frequency affects the lifetime.

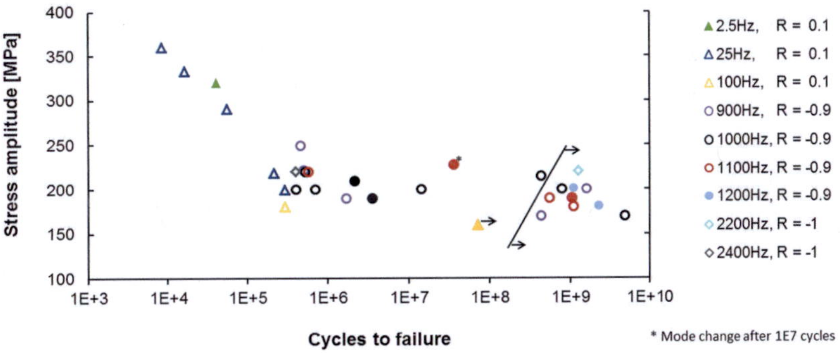

Figure 4.5. Lifetime diagram: open symbols as-cast samples, filled symbols polished samples, circular symbols acquired with resonant setup, triangular symbols acquired with setup 1 (setup with load cell). All data points located on the right from the black line which is marked by two arrows, are run outs. The data point on the left from the line which is marked with an arrow is also a run out.

The circular data points in this figure were found by means of the resonant micro fatigue setup (see section 3.2) at a load ratio close to $R = -1$ and at frequencies in the kHz-regime. At the transition between the low cycle and the high cycle regime the lifetimes are slightly elevated compared to the results from setup 1. This can be expected due to the different R-ratios which will be discussed below. Only one very high cycle fatigue failure ($>10^8$ cycles) was obtained for tests up to 2×10^9 cycles. No failure occurred at amplitudes below 190 MPa for $R = -1$. Some of the fatigue tests were conducted on samples polished on the front and back side. These tests are represented by filled symbols in figure 4.5. The lifetimes of polished samples are similar to the ones of as-cast samples (open symbols).

4.2.3. Microstructure

The microstructure next to the fracture surface of a micro molded aluminum bronze sample can be seen in figure 4.6. It was revealed by FIB milling. On the top of the image a platinum layer is visible, which was applied on the sample surface for the milling process. On the right hand side, the fracture surface can be seen. It continues perpendicularly to the image plane. Different matrix grains appear at different gray values, most likely due to different grain orientations. Their size is on the order of 5 µm to 15 µm. Various kinds of precipitates are visible and can be distinguished in terms of their morphology and their size. The two largest precipitates are dendritic-shaped having a circumference with a diameter of about 2 µm to 2.5 µm. Round-shaped precipitates with a diameter of just under 1 µm down to 100 nm or even smaller (see upper left region of the image) are concentrated in some regions. Some of them sit on the boundary between two neighboring matrix grains. Almost everywhere, the matrix is filled with lamellar precipitates. Their width is in the range of several 10 nm up to the order of 100 nm. The length varies from the order of the width (leading to round-shaped precipitates described as degenerate lamellar in [51]) up to about 8 µm. Inside each matrix grain, there is a preferential orientation of the lamellar precipitates. The lamellae tend to grow out of one matrix grain boundary into the grain in parallel to another grain boundary. In the two elongated matrix grains visible in the figure, the lamellae follow the direction of the longer grain boundary. In some regions the lamellae tend to separate into elliptic or round-shaped fragments of lamellar precipitates (see upper left region of the image). On the right hand side of the image, a triangular part of a grain, where the crack runs through, can be seen. The microstructure inside this grain is denser compared to the surrounding grains. Very fine dark and light appearing lines (the width of which is on the order of 10 nm) are arranged in parallel next to each other. Round-shaped precipitates and subgrains are present, two of which might be twins.

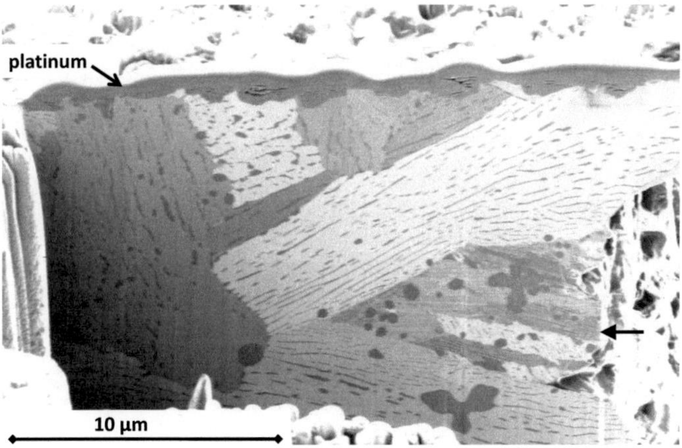

Figure 4.6. The microstructure of a fatigued micro molded aluminum bronze sample is shown in a FIB image after milling with the FIB. The arrow on the right hand side indicates a grain containing a specifically dense microstructure such as very fine dark and light appearing lines.

A further example of the microstructure revealed by FIB milling is shown in the figures 4.7 and 4.8 (the layers visible at the sample surface consist again of deposited platinum, necessary for the milling process). The cross section was prepared on a sample fatigued to 1.6×10^9 cycles without failure. For an overview, a large area is presented in figure 4.7. The matrix-grains which can be identified by the difference in contrast and by the arrangement of their containing precipitates are again of about 5 µm to 15 µm in size. A high density of precipitates is observed in the whole micrograph. Their distribution varies from one location to the other, but no precipitate-free zone which is comparable to the matrix grains in terms of size can be found. Large precipitates of about 2 µm to 4 µm in diameter are observed at the sample surface. Also lamellar precipitates grown from the surface into the bulk material are present in figure 4.7 (Right). Furthermore, clusters of point-shaped precipitates (with a size of the order of 10 nm) are visible.

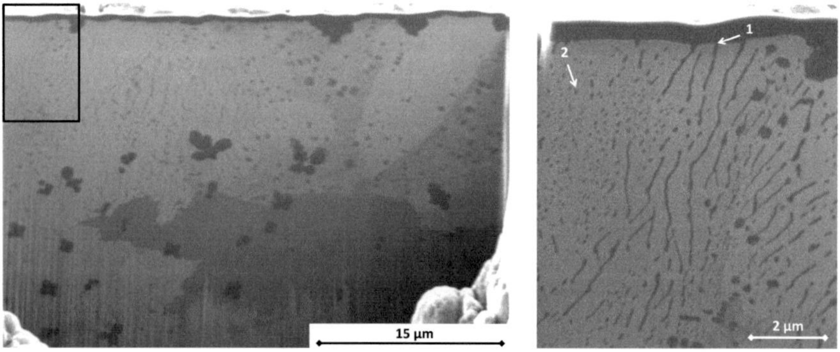

Figure 4.7. Microstructure of micro molded Aluminum bronze shown in a FIB image after milling with the FIB: (Left) An overview shows the distribution of the different kind of precipitates. The black frame indicates the region, which is magnified on the right. (Right) The magnification reveals lamellar precipitates grown from the sample surface into the bulk material (arrow 1) and fine globular precipitates similar to κ_{IV}-precipitates identified by Hasan (arrow 2).

The upper right part of figure 4.7 (Left) is magnified in figure 4.8. In addition to the observations described above, areas containing fine parallel needle like stripes are visible, which appear dark and light. Bundles of them lying next to each other are oriented in different directions. At some locations the stripes seem to be more disordered and tend to overlap each other, as can be seen in figure 4.8 (Right). In the same image, one can observe that the lamellar precipitates contained in the light matrix-grain get smaller and denser, the closer they are located to the grain boundary (the distance between lamellae ranges roughly from about 300 nm to less than 100 nm).

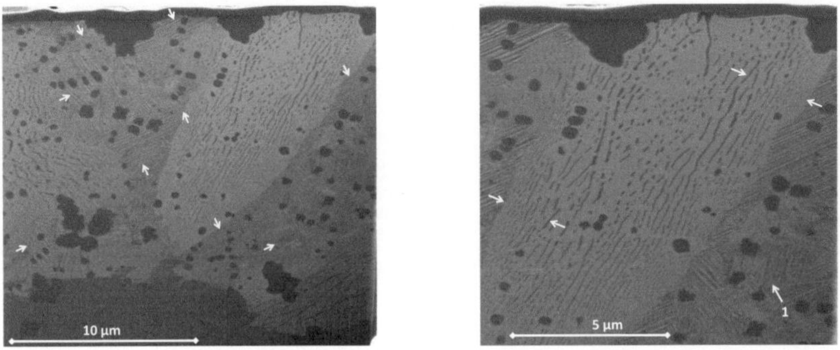

Figure 4.8. Magnification of the upper right section from the preceding figure: (Left) Areas of fine parallel needle like stripes (indicated by arrows) correspond most likely to the martensitic retained β-phase. (Right) Magnification of (Left): Needles are disordered and tend to overlap at some locations (e.g. arrow 1). Lamellar κ_{III}-precipitates in the light α-grain are smaller and arranged more densely towards the grain boundary (see areas between opposing arrows).

The material composition was investigated with EDX measurements at locations containing different kinds of precipitates. Thus, additional information is collected to identify the precipitates which are visible in the SEM images. The measurements were conducted on the etch-polished sample used for the nanoindentation tests presented above. The sites of measurement at 20 kV (mapping areas) are marked by white squares and labled (P1 to P8) in figure 4.9 (Left). The results can be grouped into three types of different compositon: regions without precipitates (matrix), regions with coarse precipitates and regions with lamellar precipitates. The corresponding portion of the elements are represented in 4.9 (Right). It can be seen that iron is mainly concentrated in the coarse precipitates. The matrix is copper-rich and contains a reduced amount of the other elements compared to the nominal composition of the alloy. The regions with lamellar precipitates show higher contents of nickel and aluminum than the matrix does. In addition to the main alloying elements, low contents of silicon and manganese were found in the alloy. The detailed results for each site are shown in figure 4.10.

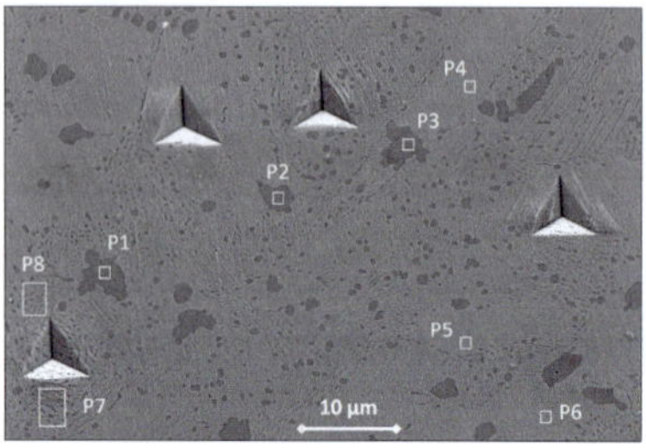

	coarse precip.	matrix	lamellar precip.
Cu	24.7 %	**87.8 %**	84.3 %
Al	**13.5 %**	7.11 %	8.5 %
Ni	**7.6 %**	3.0 %	5.0 %
Fe	**51.5 %**	1.7 %	1.8 %

Figure 4.9. EDX measurements at 20 kV reveal the material composition of regions containing different kinds of precipitates: (Left) SEM image (acquired from LEM at KIT) showing the sites of measurements P1 to P8; (Right) results are in weight percent.

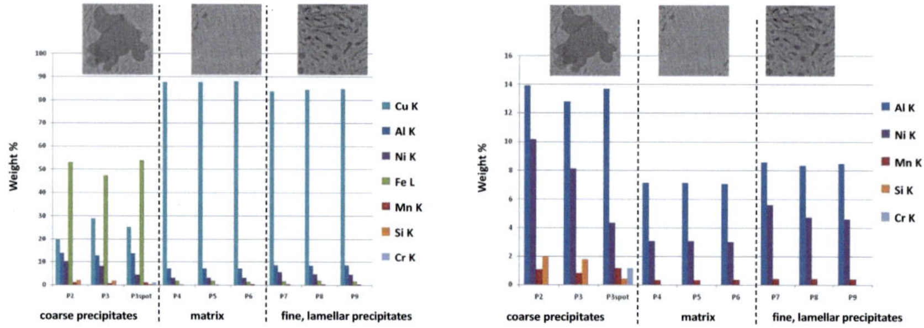

Figure 4.10. Results for EDX measurements at 20 kV shown for each site of measurement: (Left) All elements; (Right) Zoom on elements except Cu and Fe.

Additional EDX measurements were conducted at 5 kV (EDX spots). Thus, a smaller volume of material is excited lowering the signal portion coming from subsurface material, which can not be seen in the SEM image. However, a quantitative conversion of the EDX signal into the portion of contained elements is not reliable in this case, so that the data has to be interpreted qualitatively as shown in figure 4.11. Matrix regions without precipitates show a high content of copper, no iron and a small content of nickel and aluminum. The lamellar precipitates contain an elevated quantity of aluminum and nickel, some copper and small amounts of iron. The same is valid for the round-shaped precipitates of about 0.5 µm to 1 µm in size. The precipitate in figure 4.11, which was identified to be Iron-rich, is larger than the globular NiAl-based precipitates are. However, the two kinds of precipitates cannot be clearly distinguished in terms of shape or SEM contrast.

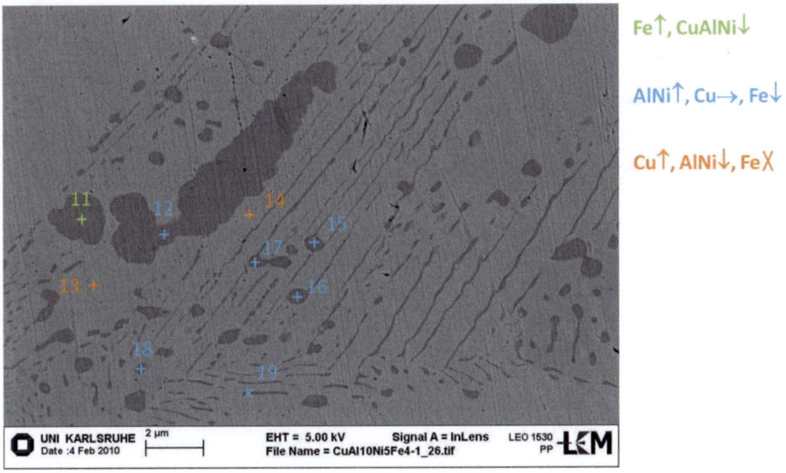

Figure 4.11. Qualitative results for EDX measurements at 5 kV (SEM image acquired from LEM at KIT): high content (arrow up), average content (horizontal arrow), low content (arrow down) and element not found (cross bar).

TEM analyses (Zeiss, LEO 1530) were conducted on the fractured sample presented in figure 4.6. A bright field image is shown in figure 4.12. Bright round-shaped precipitates, lamellar structures as well as fine dark round-shaped precipitates of sizes in the 20 nm-range are visible. The large bright globular precipitate (800 nm) could be identified as iron rich by STEM-EDX. Other bright globular precipitates of about 200 nm to 600 nm in size on this TEM lamella are rich in Ni, Al and Fe. (see 4.13) This STEM-EDX result was confirmed by EELS.

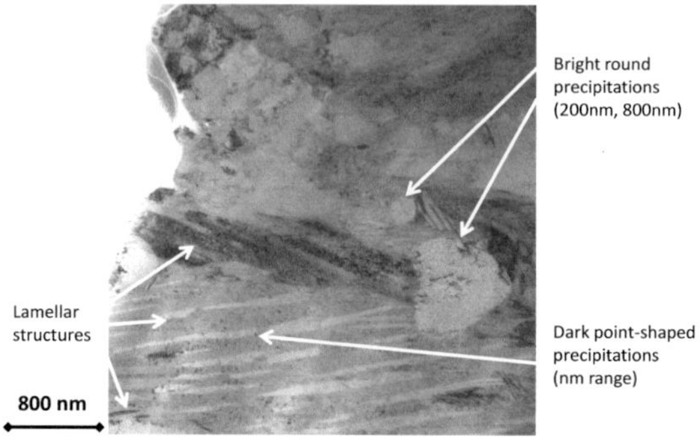

Figure 4.12. Precipitates visible in a TEM bright field image (acquired from LEM at KIT). The investigated lamella was prepared from the FIB section of the fractured sample shown in figure 4.6. The fracture surface is located on the left hand side of the TEM image.

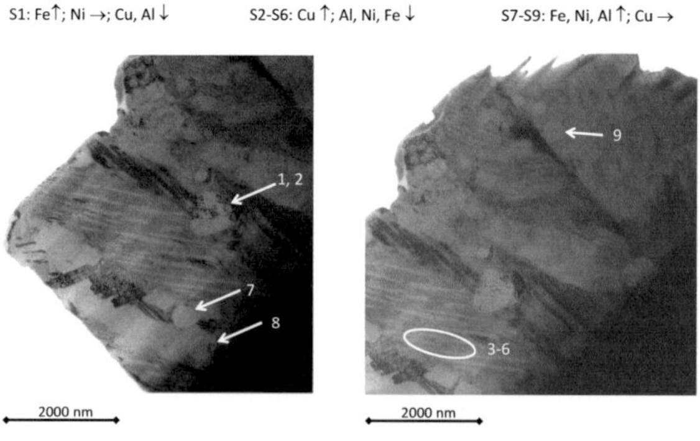

Figure 4.13. Bright field images with locations where EDX measurements were conducted (acquired from LEM at KIT). Arrows in the image indicate the locations of EDX measurements, which are numbered from 1 to 9. $S1$ is an EDX spot taken on the precipitate, $S2$ was taken next to the precipitate. The qualitative result is shown above the image. E.g. for $S1$: high content of Fe (arrow up), medium content of Ni (horizontal arrow), low content of Cu and Al (arrow down).

The fine dark round-shaped precipitates are present in a region where bright and dark gray lamellar structures can be seen. It was shown by STEM-EDX (see figures 4.13 and 4.14 a)-b)) that these bright gray and dark gray lamellae show the same chemical composition (Cu-rich with Al, Ni and Fe, see EDX locations 3 – 6 in figure 4.13). This implies that the difference in contrast comes from a difference in material structure and is not due to a different chemical composition of lamellar precipitates.

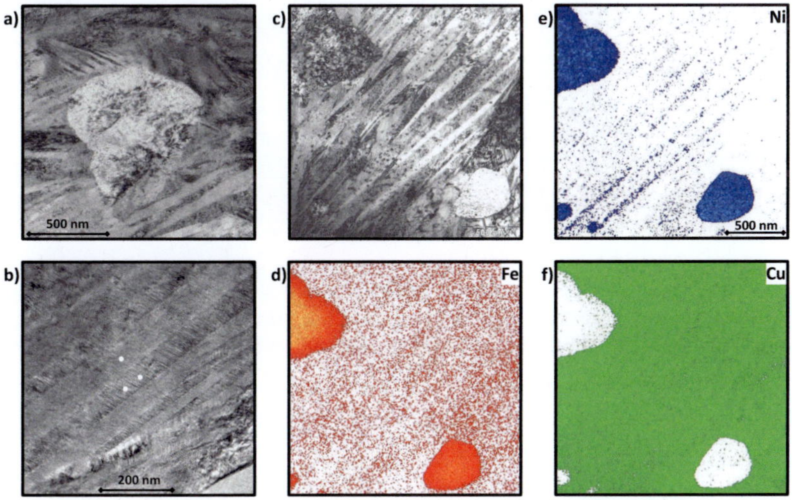

Figure 4.14. a) EDX measurements were taken at the site of the particle ($S1$ from figure 4.13) and next to the particle ($S2$ from figure 4.13), b) White dots indicate locations where EDX measurements were taken on dark and light appearing lamellae ($S3 - S6$ from figure 4.13). c) TEM image of the section where EELS-measurements were taken, d)-f) Distribution of the elements indicated in the respective figure. Areas where the element is not found appear white. For Al it was not possible to gain a suitable signal. (All images acquired from LEM at KIT.)

However, EELS from the same area indicates a trend of a slightly elevated Ni-content in lamellae which appear light in the TEM bright field image (see figure 4.14 c)-f). In this figure, a quite high noise level was accepted for Fe and Ni, in order to reveal the weak signal related to the lamellar structure. Thus, one has to interpret carefully whether the element is present or not at locations showing noise.

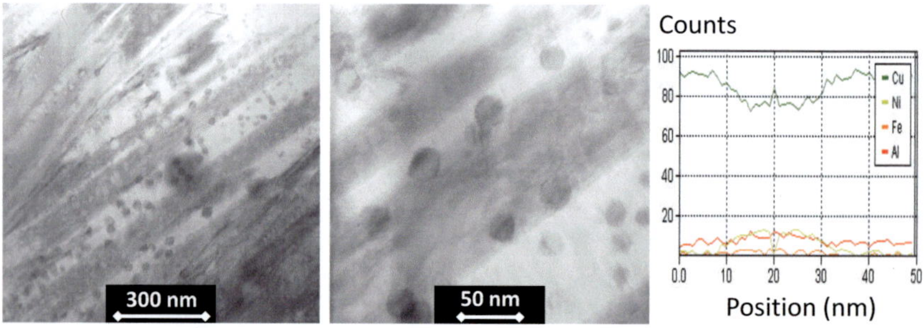

Figure 4.15. Bright field images (acquired from LEM at KIT): Small precipitates are visible in the lamellar structure. A STEM-EDX profile through one of the precipitates (20 nm in diameter) is shown in the diagram. The Al- and Ni-signals rise in between the positions 10 nm and 30 nm, where the precipitate is located, at the expense of the Cu-signal.

A second TEM investigation (FEI, Titan) on another sample also showed fine circular precipitates in regions with lamellae. They were both found in light and dark appearing lamellae. This can be seen in figure 4.15. Their size is in the range of about 5 nm to 25 nm. A lamella, rich in Ni, Fe and Al, is shown in figure 4.16. It contains various precipitates of this dimensions, the larger of which appear elliptic. One elliptic precipitate was investigated by means of EDX. On this particle, the Cu-signal rises whereas the intensities for Ni, Fe and Al diminish, as presented in the diagram of figure 4.16. The width of the lamellar precipitate itself is about 200 nm.

Many pores are visible on the side surface of the sample, sometimes containing agglomerations of material. The surrounding microstructure was

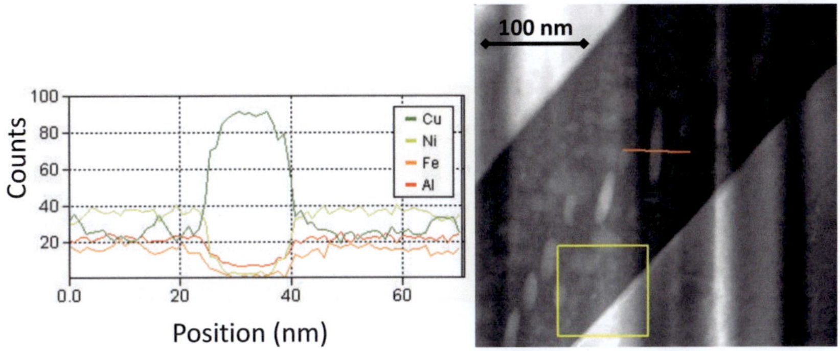

Figure 4.16. Cu-rich precipitate in lamellar Ni-rich precipitate (acquired from LEM at KIT). (The profile was measured along the line indicated in the dark field TEM image. The square illustrates the region which was used for drift correction during the measurement. Vertical lines between darker and brighter regions are due to the preparation of the lamella, the surface of which is slightly undulated).

revealed by light milling with the FIB and is presented in figure 4.17. The pores are often surrounded by a tiny layer, which shows the same contrast in the SEM as the precipitates do (indicated by arrows in the figure).

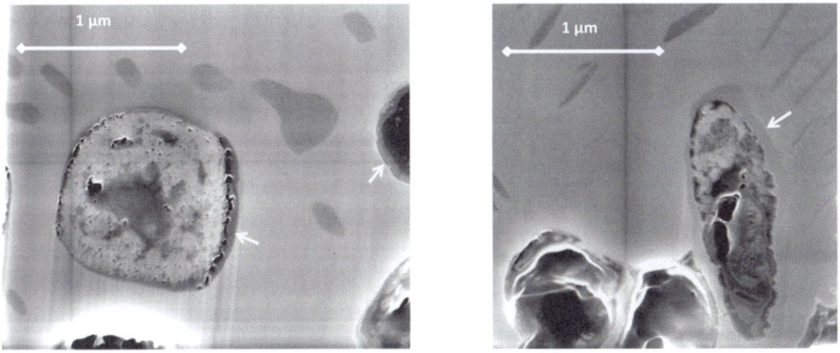

Figure 4.17. Light FIB milling at the side surface of the sample reveals a layer around surface pores indicated by arrows.

4.2.4. Damage analysis

The damage morphology of fatigued micro molded aluminum bronze is various due to numerous precipitates, which affect the fracture behavior. In this chapter, the morphology of fracture surfaces from fatigued as-cast samples will be presented at first. Second, as-cast samples containing fatigue induced microcracks will be shown, which are not separated into two parts. Therefore, the interaction of cracks and microstructural features can be revealed. Third, the damage formation during fatigue is reported for a polished sample, where the surface can be better observed. Microcracks form at three different locations before one of them becomes critical and the sample is separated into two parts. In the fourth part of this section, non fractured polished samples are presented, which contain microcracks just before catastrophic failure. This gives a deeper insight into crack path morphologies as well as surface changes in the vicinity of cracks, such as different kinds of extrusions, slip lines or secondary cracks. In the last part of this section, run outs are shown, which formed curled extrusions at surface defects. Furthermore, selective corrosion of a polished sample during fatigue is presented.

Fracture surfaces of as-cast samples

A sample that fractured after 5×10^5 cycles (load amplitude 220 MPa, load ratio $R \approx -1$) is shown in figure 4.18. The fracture surface has a pronounced surface profile and looks heterogeneous. The edge between sample surface and fracture surface is corrugated in the left part of the image, typically observed for long fatigue cracks in the aluminum-bronze-samples tested in this work. Furthermore, the crack starts at an angle of roughly 45° with respect to the loading direction, which has often been seen for these samples. Looking at the sample surface it is obvious that the roughness is not negligible compared to the sample dimensions. Since the surface is as-cast, the quality is limited by the casting process. In the presented example the site of fracture is in the transition region between gage and gripping surface.

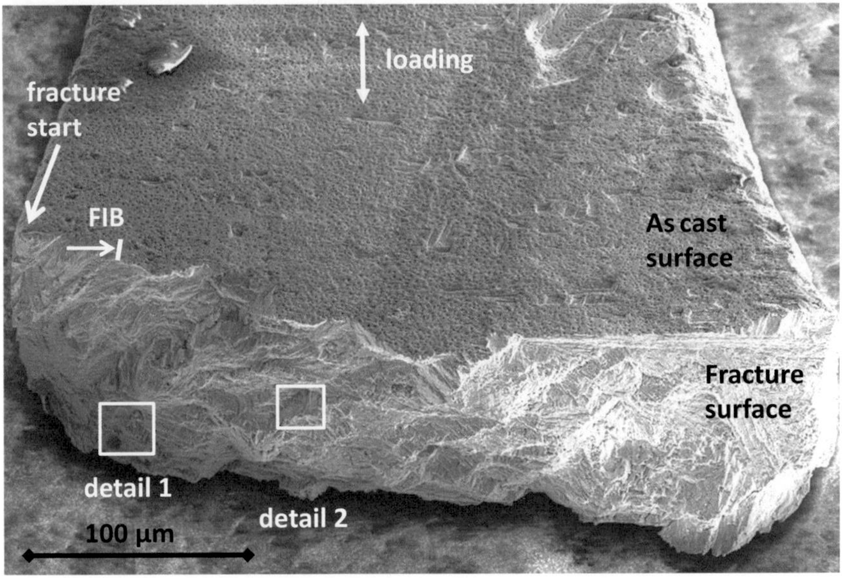

Figure 4.18. Fracture surface of a sample fatigued at 900 Hz (load amplitude $\sigma_a = 220\,\text{MPa}$, cycles to failure $N = 5 \times 10^5$, load ratio $R = -0.9$). The line labeled FIB indicates the location where the FIB cut represented in figure 4.6 was made. The arrow labeled FIB is perpendicular to the (FIB)-cutting plane and points into the direction from which the FIB cut can be observed in figure 4.6. Regions from the fracture surface, which are shown in detail in figure 4.19 are marked by white rectangles.

The location of the FIB cut, which was presented in figure 4.6 is indicated in figure 4.18 by a line and an arrow pointing perpendicularly to the cutting surface. Looking back at figure 4.6, intercrystalline and transcrystalline fracture can be observed.

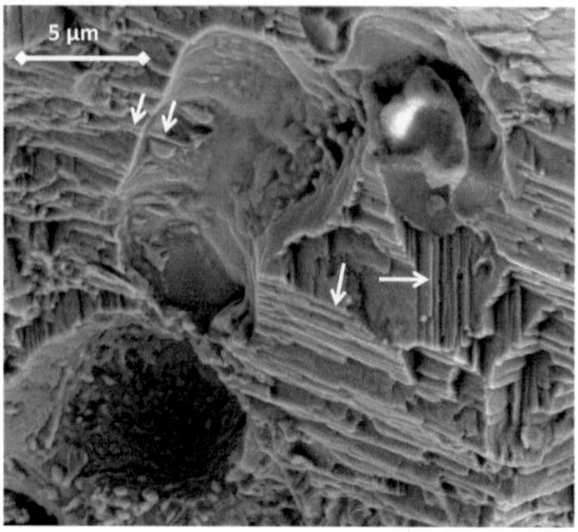

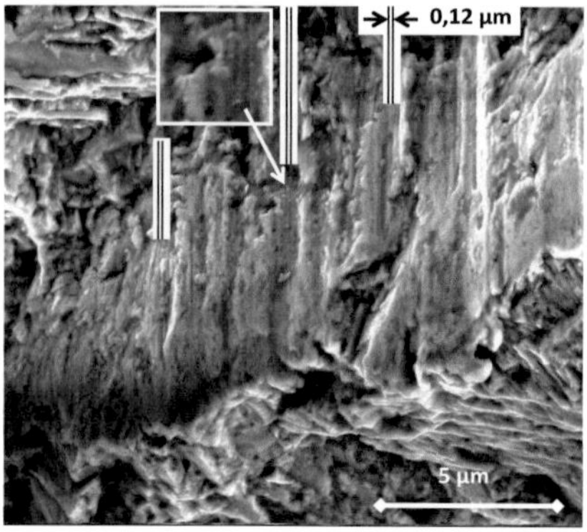

Figure 4.19. SEM images of details on the fracture surface shown in figure 4.18: (Left) Detail 1: linear structures (see arrows) with preferential orientations. Some lines continue inside a cavity which might arise from casting (see upper left arrows). The hemispheric cavity in the lower left corner of the image has a diameter of about 6 µm. Inside this cavity bumps are visible, similar to splashes. (Right) Detail 2: on a small area (about 20 µm×5 µm), which is smooth compared to the surrounding surface, striations were discovered, having a distance of about 0.12 µm. Three locations are indicated, each by black lines drawn in prolongation of two neighboring striations. The central location marked by the arrow is magnified in the white box.

Details, such as cavities, linear step-like structures and striations of the fracture surface can be seen in figure 4.19 and are described in the caption of the figure. Further details on fracture surfaces from different samples are shown in the following figures. A cavity containing bubble-like structures and a smooth facet of about 5 µm^2 found on a surface fracture with pronounced relief are presented in figure 4.20. Another cavity having lines on its surface, which look like striations or slip lines and which are surrounded by fine precipitates or particles (size scale in the several 10 nm range) is shown in figure 4.21.

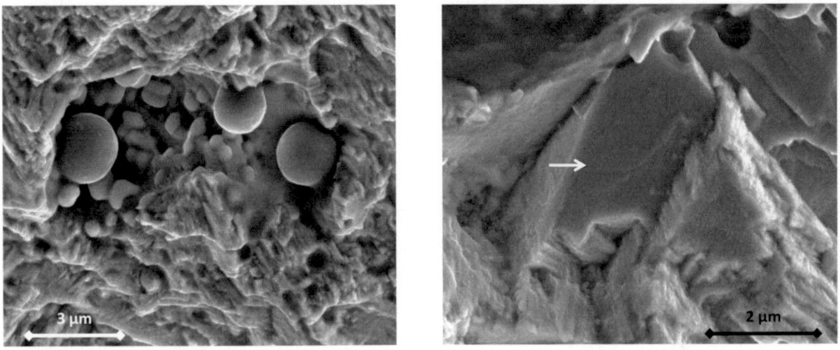

Figure 4.20. SEM images of details on the fracture surface: (Left) Bubble-like structures inside a cavity on the fracture surface (sample from figure 4.18); (Right) Smooth area on a fracture surface ($\sigma_a = 180\,\text{MPa}$, $N = 2.9 \times 10^5$, $R = 0.1$, $f = 100\,\text{Hz}$).

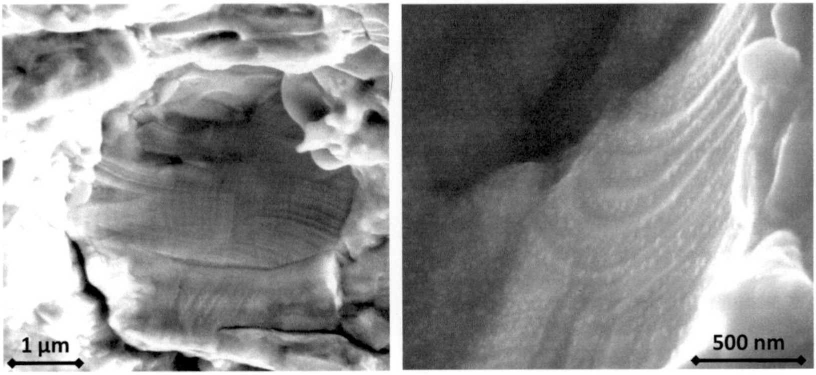

Figure 4.21. Details of a fracture surface ($\sigma_a = 180\,\text{MPa}$, $N = 2.9 \times 10^5$, $R = 0.1$, $f = 100\,\text{Hz}$): (Left) Curved lines inside a globular cavity; (Right) Zoom on curved lines showing fine precipitates in this zone.

As-cast samples containing fatigue induced microcracks

Micrographs of samples containing cracks without being separated into two parts are shown in the next figures. The cracks are barely visible on the as-cast surface and are in most cases oriented at roughly 45° with respect to the loading direction (see figure 4.22). Often, extrusions were observed on the crack (see same figure (Right)). They are denoted extrusions here, because material is extruded locally out of the sample. However, it is important to distinguish between such extrusions coming out of the cracks and purely slip induced extrusions. The latter form at non cracked areas and lead to surface roughening, which can cause crack initiation.

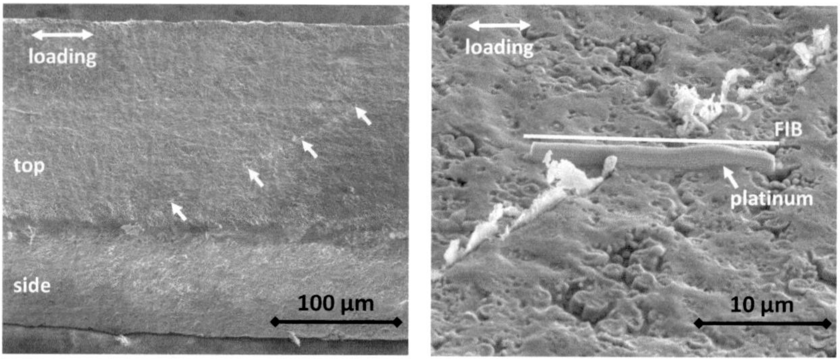

Figure 4.22. SEM images of a fatigued sample ($\sigma_a = 220\,\text{MPa}$, $N = 4 \times 10^5$, $R = -1$, $f = 2362\,\text{Hz}$), (Left) Arrows point to a detected crack. (Right) Magnified view of the crack showing extrusions coming out of the crack. Platinum was applied in order to prepare the FIB cut shown in the next figure.

FIB cuts revealing the microstructure and the crack propagation inside the samples are presented in the figures 4.23-4.25. The crack depicted in figure 4.23 (Left) is preferentially oriented parallel to surrounding lamellar precipitates. In the upper part of the image, the crack separates two regions of different material contrast indicating grains of different orientations. A subsurface precipitate of about 1 µm in diameter is located close to the crack. Deeper inside the material, the crack kinks in front of a material portion of darker contrast surrounded by two lamellae oriented into the direction of the crack. In the following, the crack propagates again in parallel to lamellae slightly inclined to the first direction, until it penetrates a region of darker material contrast. There, the lamellar precipitates are oriented nearly perpendicular to the lamellae mentioned before, and are broken by the crack. However, the crack is deflected again and the angle between crack and lamellae is roughly 45° (note, the value is only a rough estimate to point out that it is neither close to 0° nor close to 90°; the image was taken at a stage tilt of 52°). The further propagation is magnified in figure 4.23 (Right). The crack follows in the direction of a lamellar precipitate, as soon as it encounters one that is oriented in a similar direction as the lamellae the crack followed initially. In the magnified view it looks like the crack would propagate very close to but not inside the lamellar precipitate.

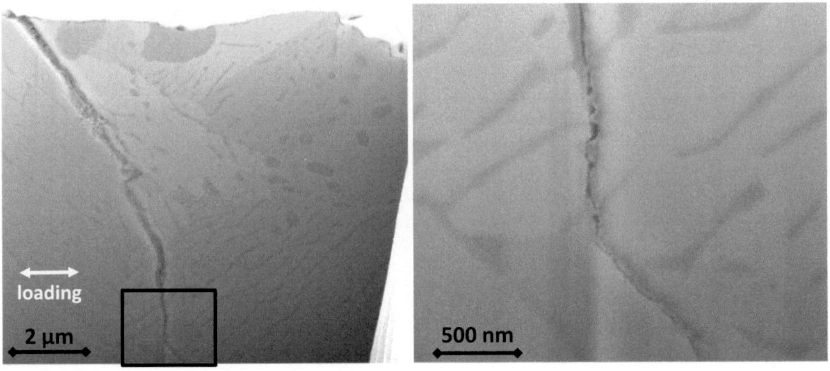

Figure 4.23. The crack propagates preferentially in parallel to lamellar precipitates. (Left) Next to a subsurface precipitate the crack penetrates into the material between two regions of different brightness. It kinks in front of a darker appearing region. The black box is magnified in the figure (Right).

This observation is supported by figure 4.24. In the image on the left hand side, an oblique view of the sample shows a crack propagating at the top and at the side surface of the sample. From the FIB cuts of this sample, the angle between the loading direction of the sample and the normal of the crack surface was determined to be 47±2 °. Cut number 1 was milled in parallel to the side surface of the sample at the location indicated by the horizontal line denoted FIB 1. The revealed microstructure is shown in 4.24 (Right). A straight crack path along lamellar precipitates is clearly visible.

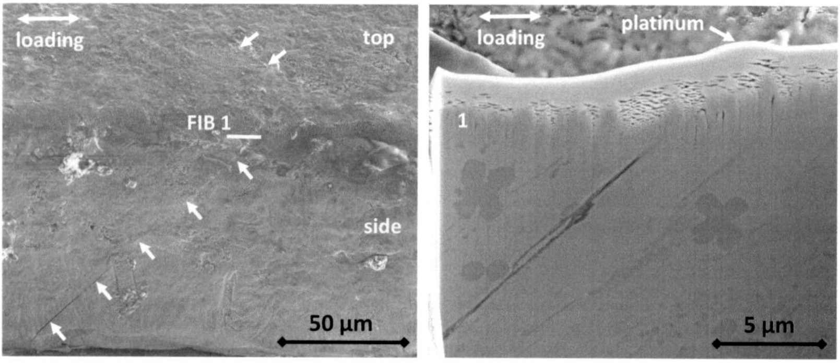

Figure 4.24. Fatigued as-cast sample containing cracks ($\sigma_a = 200\,\text{MPa}$, $N = 1.4 \times 10^7$, $R = -0.9$, $f = 1035\,\text{Hz}$); (Left) Oblique view at 52° on the sample side. A crack indicated by arrows appears on the side surface (close to 45° with respect to the loading direction) and continues on the top surface. The line FIB1 marks the location of a FIB cut milled in parallel to the side surface. The result is shown in the figure (Right): The crack propagates in parallel to / along lamellar precipitates.

A second FIB cut was milled at 90° with respect to the side surface of the sample, and is shown in figure 4.25 (Left). Also in this case, the crack propagates along lamellar precipitates. Furthermore, the crack is not straight but slightly wavy, similar to the surrounding lamellar precipitates. The top surface of the sample containing the two FIB cuts is shown in the oblique view of figure 4.25 (Right). In the black box, a detail of the image is magnified, showing the same kind of extrusions pressed out of the crack on the sample surface, which were found on the sample presented in figure 4.22.

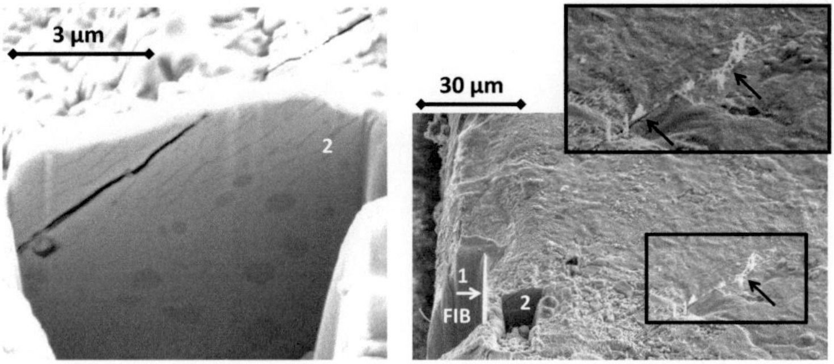

Figure 4.25. Oblique views (52° tilt): (Left) FIB cut milled at 90° with respect to the cut shown in figure 4.24. Also in this direction the crack runs in parallel to / along lamellar precipitates. (Right) Oblique view in the direction of loading on the top surface (roughened by the ion beam). In the black box a part of a crack is magnified, which runs in parallel to the crack revealed by FIB milling. Extrusions that emerge from the crack are visible (indicated by arrows).

The side surface of the sample shown in figure 4.24 (Left) was slightly polished by FIB milling in order to reveal the microstructure. The result is presented in figure 4.26. The large straight crack and a smaller secondary crack are visible. Obviously, the secondary crack is deflected at locations where it meets round-shaped precipitates. The main crack is large enough to maintain its initial propagation direction along the precipitates. However, it does not penetrate through the precipitates, but it is locally deflected along the interface between matrix and precipitate. Both cracks are mainly oriented in parallel to lamellar precipitates. Kinking of the secondary crack is observed at multiple locations. At two locations, straight parts of the crack having a length in the 200 nm range propagate orthogonally to the direction of lamellar precipitates. Before and after these sections, the crack propagates along the main propagation direction, i.e. parallel to the lamellar precipitates.

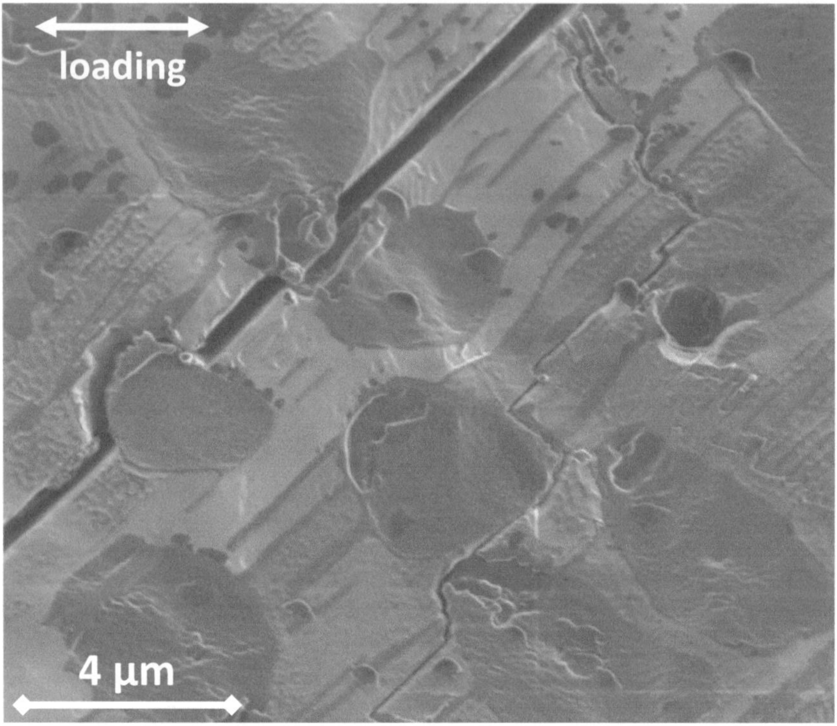

Figure 4.26. Magnified section of the crack on the side surface, which was shown in figure 4.24. The crack is deviated by round-shaped precipitates and propagates preferentially along lamellar precipitates. Crack kinks are observed on the thin secondary crack.

A TEM lamella containing the main crack was cut out next to the section shown in figure 4.26. The dashed rectangle in figure 4.27 (Left) indicates the location from which the lamella was lifted out; (the TEM investigations shown in the figures 4.15 and 4.16 were conducted on the same TEM lamella). The crack is indicated by an arrow in the central TEM bright field image of the figure. A lamellar precipitate with light contrast is located about 2 µm apart and oriented in parallel to the crack. The continuation of the crack is shown in figure 4.27 (Right). In this section the crack has about the same thickness as the lamellar precipitate mentioned before (on the order of 200 nm).

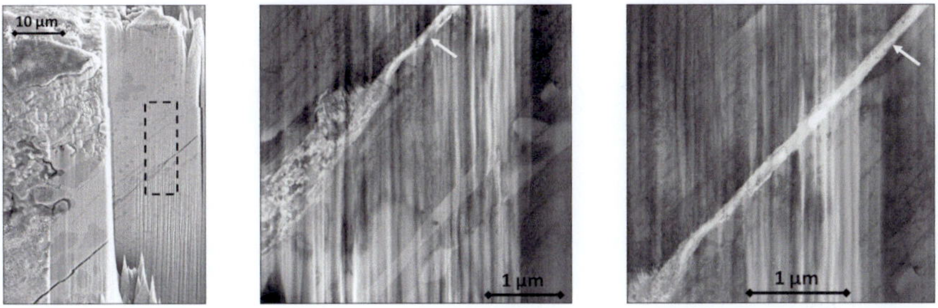

Figure 4.27. Crack investigated by means of TEM (acquired from LEM at KIT): (Left) SEM image of the region where the lamella was lift out. The dashed rectangle indicates the estimated location of the TEM lamella. (Center) and (Right) STEM bright field images of the crack - arrows show the crack. (Center) A lamellar structure is visible, which is oriented in parallel to the crack.

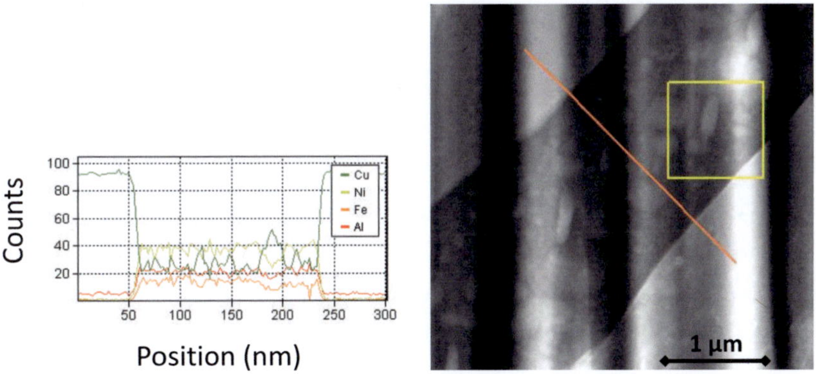

Figure 4.28. EDX profile of lamellar precipitate oriented in parallel to the crack (acquired from LEM at KIT): The lamella is rich in Ni, Al and Fe compared to the surrounding matrix material. (The profile was measured along the line indicated in the TEM image. The square illustrates the region which was used for drift correction during the measurement.)

An EDX profile of a 200 nm large lamellar precipitate oriented in parallel to the crack is presented in figure 4.28. The diagram represents the number of counts detected over the position, which means that the content of aluminum with its lower atomic number is elevated in comparison to the content of iron, copper and nickel, at locations where all elements show a similar number of counts. At the position of 190 nm the signal of Cu increases, whereas the content of the other elements decreases. This can be correlated to the position of a precipitate with bright contrast in the TEM image.

Fatigue damage formation of a polished sample at three locations before failure

The following images come from samples which were polished before the fatigue tests. Images captured during the experiment with the camera are presented in figure 4.29. The horizontal view of the whole sample at the bottom of the figure contains boxes marking the details, which are magnified above and shown after different numbers of cycles in the form of a table. A first crack can be seen at location 1 after 1.707×10^6 cycles (first line). It has grown after 2.786×10^6 cycles (second line) to a size where it can be seen in the image (marked by an arrow). After 1.718×10^7 cycles a second crack appeared (see arrow in line 3) at location 2 and the crack at location 1 has grown. Just before failure (3.568×10^7 cycles), a third crack appeared at location 3. The fractured sample after 3.613×10^7 cycles is represented on the right hand side of the figure. It was separated at location 2. After 1.2×10^7 cycles, an increasing noise level was detected in the mass displacement signal (i.e. in the load signal). This was most likely related to the damage formation at location 2.

SEM images of the cracks which had formed at locations 1 and 3 are presented in the following figures.

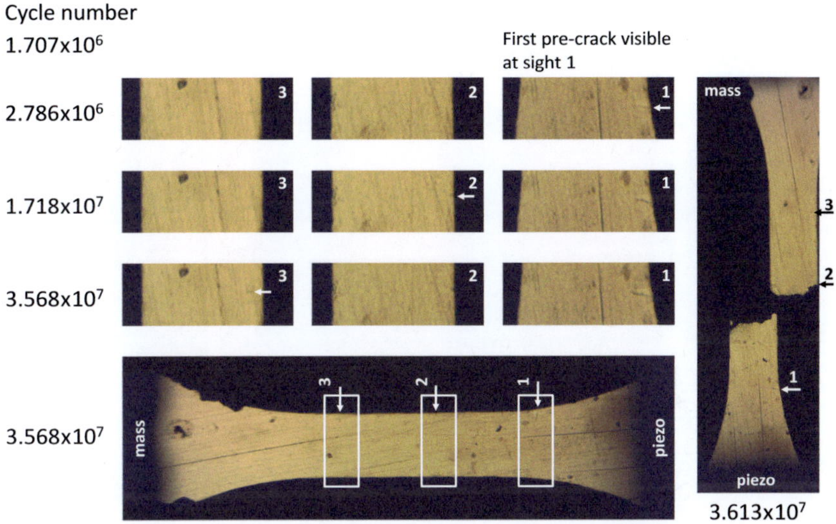

Figure 4.29. Damage formation observed during fatigue of a polished sample (σ_a = 228 MPa, $N = 3.6 \times 10^7$, R =-0.9, f = 1142 Hz). Three zones of damage are magnified after different numbers of cycles. Arrows indicate the locations of damage. First damage occurs at location 1 after 1.7×10^6 cycles. A small crack appears at location 2 after 1.7×10^7 cycles. The sample failed after 3.61×10^7 cycles (represented on the right hand side).

In the central image of figure 4.30, the whole crack is presented (observed on the same polished sample surface which is shown in figure 4.29). The black box on the top of this image is magnified at the bottom and contains a feature at the side surface of the sample, which is indicated by an arrow. This curly feature is suggested being related to the fatigue experiment, since similar observations were made on other fatigued samples. EDX mapping in an SEM on a similar feature returned signals for all elements contained in the alloy at portions similar to the matrix material. The crack path is corrugated. In the magnification on the right hand side, kinking is observed. On the left hand side, the part close to the crack tip is magnified. The changes in direction are more pronounced and the crack meanders. An extrusion coming out of the crack appears as a white dot. A higher magnification of the crack tip is presented in figure 4.31. Film like extrusions, which are thin and laminar are observed on the crack.

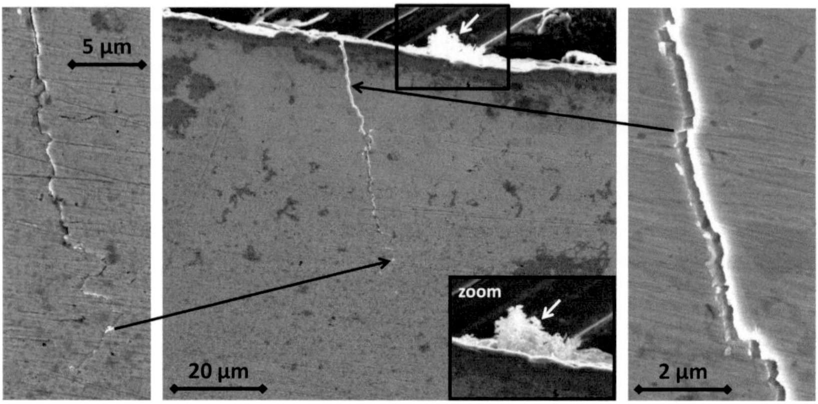

Figure 4.30. The crack of location 1 from figure 4.29 is shown in SEM images. On the right and on the left hand side, the crack is partially magnified. Furthermore, a curling feature is magnified in a zoom insert on the central image. Similar features have been observed on other fatigued samples and are probably related to damage.

Figure 4.31. Tip of the crack shown in figure 4.30 imaged by SEM. Thin, film like extrusions grow out of the crack (the loading direction is horizontal).

The same kind of observation was made at location 3 (see figure 4.32 (Left)). The whole crack is magnified in 4.32 (Center). The central part of the crack is quite straight, whereas the upper part shows branching and the lower part shows again film like extrusions. On the right hand side the upper part of the crack showing branching is magnified. Here, the extrusions are more like chips than film like.

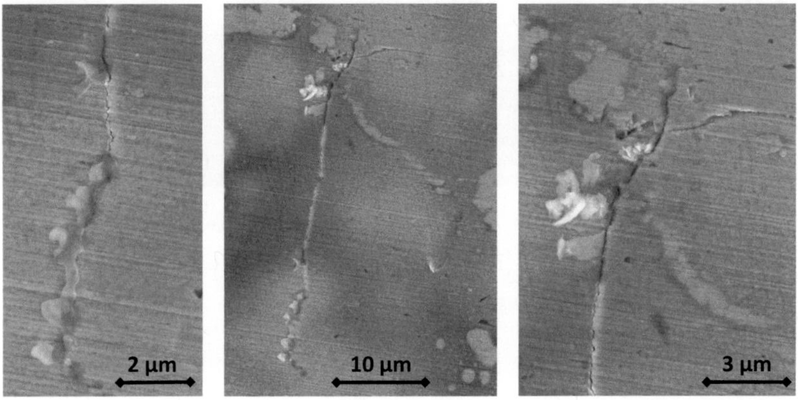

Figure 4.32. The whole crack of location 3 from figure 4.29 is shown in the central SEM image (the loading direction is horizontal). The upper part showing crack branching and extrusions is magnified on the right. The lower part with thin, film like extrusions is magnified on the left.

The fracture at location 2 is shown in figure 4.33. In the image (Left), three secondary cracks on the sample surface are marked with arrows. The upper two of them are oriented at an angle very close to 45° with respect to the loading direction. The same is valid for the very straight part of the fracture edge at the top right of the image (this is the location where the damage before fracture could be observed in the camera during the experiment). The secondary crack at the bottom of the image has a larger angle with respect to the loading direction which however is smaller than 90°. In the figure (Right), the straight part of the fracture surface is presented in an oblique view along the loading direction. The upper two secondary cracks from the image (Left) are marked again with arrows. Obviously, not only

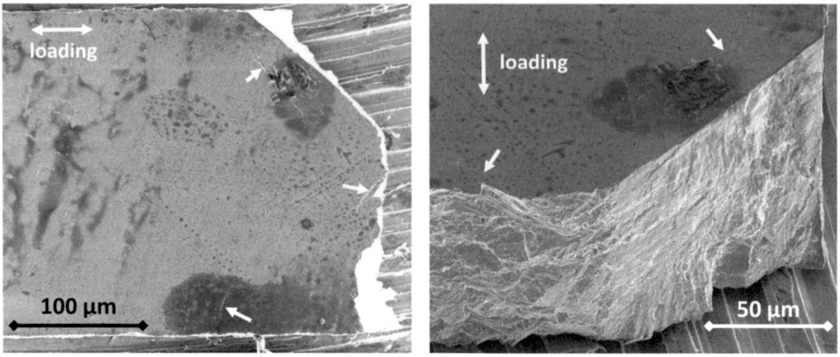

Figure 4.33. Fracture at location 2 from figure 4.29: (Left) Three secondary cracks are marked by arrows. Two of them are oriented at 45° with respect to the loading direction. The same is valid for the straight part of the fracture path (top right). (Right) Oblique view on the straight part of the fracture surface in loading direction.

the fracture edge at the surface, but the whole part of the fracture surface is straight and relatively smooth.

The sample was rotated about 45° in order to face the straight part of the fracture surface. The view is presented in figure 4.34. In the figure (Left), the side surface of the sample is visible at the very right border. The neighboring fracture surface shows a very flat area containing horizontal lines and gets granularly structured to the left. In the upper part of the image, the polished fracture surface is observed. A dark contrast is visible along the edge between fracture surface and sample surface. In magnified views, it looked like a very thin bulge or film. In the figure (Right), a magnification shows that the horizontal lines on the straight fracture surface are related to steps of plate like structures (indicated by arrows). It is interesting to compare this observation to the linear structures shown in figure 4.19. The diameters of the granular fragments on the fracture surface are on the order of 1 µm in size.

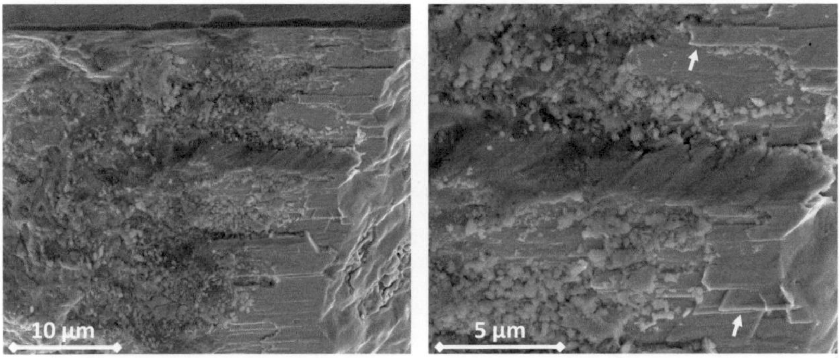

Figure 4.34. Front view of the straight section of the fracture surface: (Left) Sample side surface at the very right border of the image. The adjacent fracture surface is smooth containing horizontal lines and gets granular to the left. (Right) Transition between the smooth surface containing steps (indicated by arrows) and the granular surface.

The central secondary crack of figure 4.33 (Left) is presented in figure 4.35, in an oblique view on the fracture surface and marked by arrows. On the fracture surface, edges parallel to this crack can be observed. The crack is clearly visible on the sample surface and continues on the fracture surface, as magnified in figure 4.35 (Right). Here, kinking of the crack is observed, similar to the examples shown earlier, such as figure 4.26. Further microcracks can be observed on the fracture surface in this image.

The tip of this microcrack on the sample surface is magnified in figure 4.36 and reveals further microcracks in the vicinity. A zig zag path of the main secondary crack is observed. The microcrack on the left is partially oriented in parallel and shows the same kind of kink. At the upper right corner of the image, a microcrack starting from a corner of the fracture surface follows the same direction as the main secondary crack. The arrows in the image mark extrusions coming out of microcracks. These extrusions indicate the paths of microcracks which cannot be seen themselves in the image. Some of these extrusions are magnified in 4.36 (Right). They appear very locally

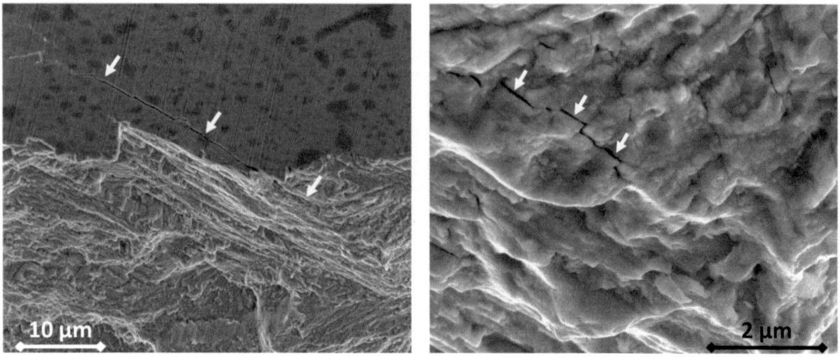

Figure 4.35. Oblique view in the loading direction: Secondary crack from figure 4.33 (marked by arrows) continues on the fracture surface. The magnification of the section on the fracture surface (Right) shows kinking and further microcracks in the vicinity.

and look very bright in the SEM image. Their shape is bubble like and they were predominantly observed at locations where the crack changes in direction or stops.

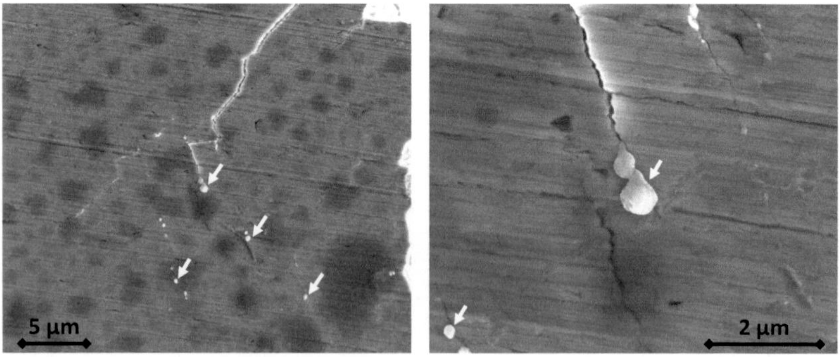

Figure 4.36. The loading direction is horizontal: (Left) The crack tip on the polished sample surface shown in figure 4.35. Secondary cracks are observed (some of them by extrusions indicated by arrows). They are oriented in parallel to segments of the main crack. (Right) Magnification of extrusions coming out of the cracks, mainly at locations where the crack is stopped or deviated.

Polished samples containing fatigue cracks just before catastrophic failure

A crack traversing about 250 µm of the sample's width (at the transition between gage and gripping surface) is shown in figure 4.37. This crack runs through the whole side surface at an angle of about 40° with respect to the loading direction (horizontal) taking into account the tilt of the stage in the microscope (45°) but not the curved side surface in the transition between gage and gripping surface. Thus, the angle between crack and loading direction is larger than 40° when projected on a plane parallel to the gage side surface.

A view of the sample top surface containing the whole crack is presented in the second image from the left. Starting from the bottom of this image (edge of the sample side), the crack propagates at first at an angle of about 45° with respect to the loading direction (horizontal). Having reached a length of about 80 µm, the crack changes direction and continues after a transient part of about 20 µm to 30 µm towards an overall propagation direction, which is perpendicular to the loading direction with a deviation of ±10°.

Sections of the crack are magnified in the two images on the right hand side (the arrows between pictures connect identical locations shown in different images). It is obvious that the crack profile is strongly corrugated and the crack often changes locally in direction.

In the very right image of figure 4.37, three details are indicated by arrows. Detail 1 shows bright appearing cauliflower-like depositions which were also observed on some as-cast samples. EDX measurements on an as-cast sample at locations containing these structures showed elevated Ni-contents compared to regions without these structures. FIB cutting revealed that the structures were not everywhere in contact to the sample surface. The suggestion that the features might be related to pores on the sample surface could not be confirmed. Detail 2 shows a dark appearing area close to the crack tip. Similar observations were made on other polished samples (see some pictures shown earlier). In some cases these locations could be correlated to locations where a brightness change was observed during the experiment with the camera. In the present case, the dark area is centered around the crack.

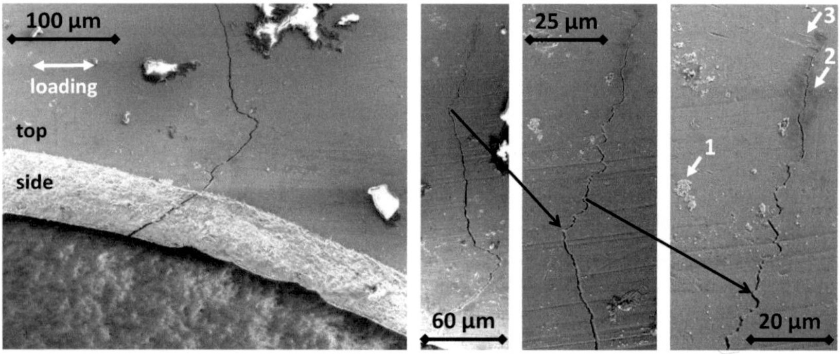

Figure 4.37. A crack runs at the transition between gage section and gripping surface from the side surface into the sample (σ_a = 190 MPa, $N = 3.5 \times 10^6$, $R = -0.9$, f = 1005 Hz). (Left) Oblique view at 45° tilt: the crack propagates on the side surface at an angle between 40 and 45° with respect to the loading direction. (Center-Left) Top view of the polished surface containing the crack; (Center-Right) Magnified view: The black arrows mark identic positions in neighboring pictures. (Right) Magnified view: Arrows indicate (1) a cauliflower-like deposition which was also observed on some fatigued as-cast samples, (2) a dark zone on the sample surface which appeared also on other samples and is noticeably concentrated around the crack here and (3) glide steps with extrusions at the crack tip.

Detail 3 marks glide steps and extrusions at the end of the crack tip, which are magnified in figure 4.38 (Left). They are oriented roughly at 90° with respect to the direction of the arriving crack. Three crack branches are oriented in the direction of the glide bands which prolongate the latter and are arranged in between them. At the top of the image three arrows mark (from the left to the right) a microcrack prolongating the extrusions, the highest visible extrusion and a glide band visible as dark line. The two arrows in the lower part of the image mark dark parallel lines which show the same brightness as the line mentioned before and which are oriented in a different direction.

The crack section close to the sample edge, which is oriented at 45° with respect to the loading direction is partially magnified in figure 4.38 (Right). Here, the crack is widely open. On the crack surface, a cavity with a diameter in the range of 1 µm is visible. Furthermore, steps related to the kinking of the crack are observed, and the crack surface appears smooth. On the sample surface, white arrows mark fine dark appearing lines, which are oriented in parallel and perpendicularly to the crack.

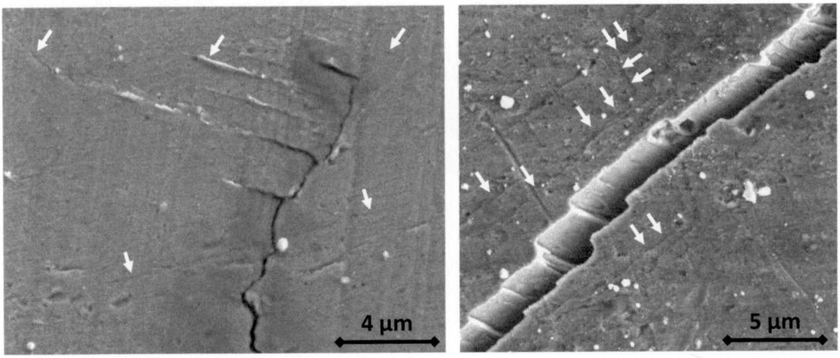

Figure 4.38. Sections of the crack shown in figure 4.37 in magnified views: (Left) Oblique view (45°tilt) of the extrusions and of dark lines at the crack tip, indicated by arrows; (Right) Widely opened section of the crack close to the sample edge showing steps on the fracture surface and dark lines on the sample surface (marked by arrows).

Often, secondary cracks were observed besides the largest main crack. In addition to cases shown earlier, some examples are presented in figure 4.39 (loading direction horizontal). In the figure (Left), a secondary crack propagating in parallel to the main crack shown in figure 4.37 is visible. The image in the center shows another sample containing a main crack inside the gage section, which splits into two branches (marked by two arrows at the top of the image). This crack is stopped at the bottom of the image in favor of a parallel crack (about 10 µm apart), which reaches the edge of the sample. Secondary cracks are marked with arrows. One of them is close to the main crack (barely 10 µm apart), the others are concentrated in the upper right part of the image. The letters a, b and c in figure 4.39 (Center) mark cracks, the orientations of which are very similar.

Figure 4.39. Secondary cracks, which are oriented in parallel to the main crack (loading direction horizontal): (Left) Sample from figure 4.37; (Center) Fatigued polished sample with microcracks oriented in parallel to sections of the main crack; Similar orientations are designated by the same letter (2.1×10^6 cycles, 210 MPa, $f = 1007$ Hz, $R = -0.9$). (Right) oblique view on the main crack and a secondary crack reaching the side surface.

The change from the stopped main crack to the crack reaching the edge of the sample is magnified in the image (Right). The oblique view shows the side surface and the top surface of the sample. The crack on the left runs through a surface dent coming from the casting process (see also figure 4.39 (Center)). The crack on the right (main crack) does not reach the side surface of the sample. The tip of the branching main crack shown in figure 4.39 (Center) is magnified in figure 4.40. The image is rotated about 90°.

Thus, the loading direction is vertical. The main crack is visible in the upper right image of the figure. It propagates from the right to the left and the location where it branches is indicated by a white arrow. The small branch which stops after a short distance (nearly 10 µm) is visible in the form of a dark line. This part of the crack is not open. The upper right area in the image contains a highly undulated crack surface section of the main crack, which is magnified in the black box below.

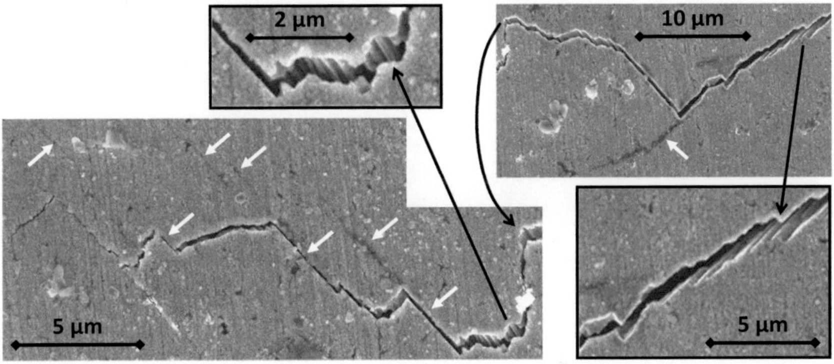

Figure 4.40. Magnification of the crack tip from the main crack shown in 4.39 (Center); The view is rotated about 90°, thus the loading direction is vertical. In the top right image, branching of the crack is observed marked by the white arrow. A corrugated crack section is magnified in the black box below. The continuation of the crack is presented in the lower left image. Another corrugated crack section is magnified in the black box above. White arrows indicate dark lines which are oriented in parallel to crack sections or prolong the latter.

The continuation of the main crack toward its tip (left border of the upper right image) is shown in the lower left image (figure 4.40). Again, an undulated section is magnified (see black box above). It switches to a rather straight propagation mode as soon as it changes direction toward an orientation of 45° with respect to the loading direction. In the lower left image, one can observe that the crack shows many segments which are oriented at 45° with respect to the loading direction. The longest three of them are between 3 and 4 µm long. By trend, these segments are opened less than segments which are not oriented at 45°. White arrows in the image indicate dark lines, which prolongate the 45°-segments or which are oriented in parallel to them.

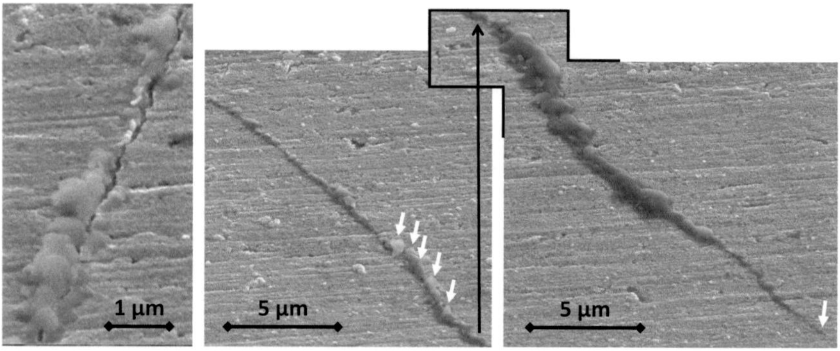

Figure 4.41. Magnification of the secondary cracks shown in figure 4.39 (Center). (Left) The crack is partially overgrown by material. (Center) and (Right) A bulge covers the whole crack and looks like a sequence of extrusions (see sequence of arrows in the central image). At the end of the bulge, a black line is visible on the surface (indicated by the arrow in the image (Right)).

The two secondary cracks, which are marked in the upper right corner of image 4.39 (Center) with the orientations a and c are magnified in figure 4.41. One crack is partially overgrown with material, as can be seen in figure 4.41 (Left) (showing the crack with orientation c). The other crack (orientation a) is entirely covered by a bulge of material (see images (Center) and (Right) in the figure). The morphology of the bulge looks like a sequence of extrusions shown in 4.36, however the material contrast is completely different. The bulge is not as bright as the extrusions, but it shows a contrast similar to the sample surface. At the very end of the bulge (see arrow in figure 4.41 (Right)) a dark line is visible, similar to the lines described in figure 4.40.

Extrusions on run outs and corrosion during fatigue

Quite long extrusions emerge on the crack and form curling structures (figure 4.42). In the figure (Left) two extrusions have grown out of a crack on the polished sample surface and are marked by arrows. They take on the shape of a plate rather than of a sphere but show mainly the same material contrast as the extrusions from figure 4.36 do. Indeed, the upper extrusion shows partially this bright contrast and partially the dark sample contrast as it was observed for the bulge in figure 4.41. At the location where these two extrusions were found by SEM, a local contrast change was observed in the camera during the fatigue test. The bright wire like structures in the lower part of the image arise locally on the side surface of the sample (only a few similar features were found on the same sample). EDX mapping gave a material composition of these wires which corresponds to the matrix material of the aluminum bronze.

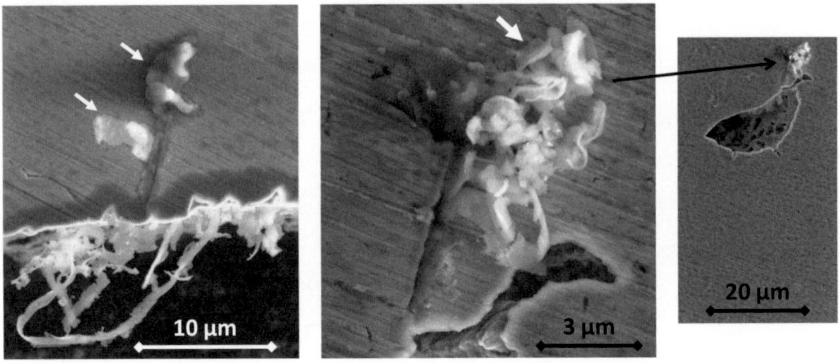

Figure 4.42. (Left) Plate like extrusions coming out of a 10 µm long crack on the polished sample surface (run out: 1.1×10^9 cycles, 190 MPa, f = 1056 Hz, $R = -0.9$). The crack starts at the side surface, where wire like structures are observed. (Center) and (Right) Curling extrusions appeared at a crack next to a casting defect (run out: 2.3×10^9 cycles, 180 MPa, f = 1202 Hz, $R = -0.9$).

Curling extrusions appearing at one spot, which were observed on another sample, are presented in figure 4.42 (Center). They are located close to a casting defect which is shown in the figure (Right). A contrast change was registered during fatigue at the same location with the camera. The curling structure (see (Center)) arises at the end of a short crack similar to the extrusion in the figure (Left). The morphology of the curling structure (see (Center)) is very similar to the morphology of curling structures which are present at the base of the wire like structures in the figure (Left). Both samples shown in figure 4.42 did not fail. They were fatigued in the 1 kHz-regime at 190 MPa and 180 MPa until they reached 1.1×10^9 and 2.3×10^9 cycles respectively.

A dark crusty layer on the polished surface of a fatigued sample was already contained in figure 4.33. This feature is magnified in figure 4.43 (Left). For further investigations, a FIB cut was performed along the double arrow in the lower right corner of the image (the location of the line is roughly estimated). The dashed line indicates that the FIB cross section continues in both directions and is not fully contained in the picture. The arrow pointing perpendicularly to the dashed line indicates the direction of view onto the FIB cross section which is presented in the figure (Right). In this cross section it was observed that the sample surface beneath the crusty feature is attacked at locations where the fine needle like microstructure is present (marked by the two oblique arrows). In regions with the ordinary matrix containing lamellar and round-shaped precipitates (light regions), no surface attack is observed. However, the crusty feature on the surface also extends into these regions (see upper left border of 4.43 (Right).

The attacked area is magnified in figure 4.44 (Left). It looks granular and consists of bright appearing particles of several 100 nm down to less than 10 nm. In some regions these particles are elongated or lamellar. The inter-

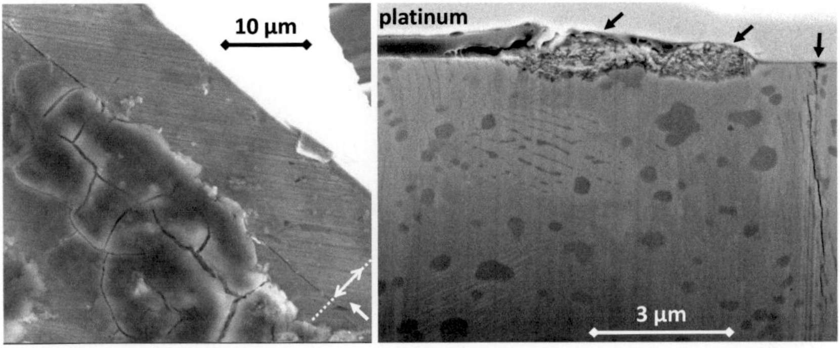

Figure 4.43. (Left) Crusty layer on the polished sample surface (fatigued sample, which was already shown in figure 4.33). The white line indicates the location of a FIB cut shown in the image (Right): Regions beneath the crusty structure where the fine needle like microstructure is present are attacked (marked by oblique arrows). A surface pore at the end of a crack is marked by an arrow.

face between the attacked material and needle like microstructure looks like it has been affected by pitting. Figure 4.44 (Right) shows another cutting plane at the same location. The arrow marks a round-shaped precipitate which is not attacked but which is partially surrounded by the attacked needle like microstructure.

Crusty features which were found on a fatigued as-cast sample and which resemble the feature presented before are shown in figure 4.45. Small particles (several 100 nm and smaller in size) are observed inside the semi-electron-transparent crusty feature in the image (Right).

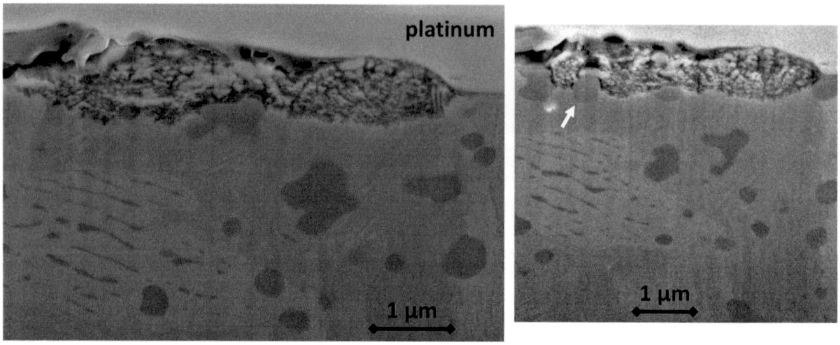

Figure 4.44. The attacked region shown in figure 4.43 is magnified in the image (Left). The attacked material looks granular and the interface to the needle-like microstructure looks pitted. The round-shaped precipitates are not attacked. This is also confirmed by another cutting slice at the same location in the image (Right) (see precipitate marked by arrow).

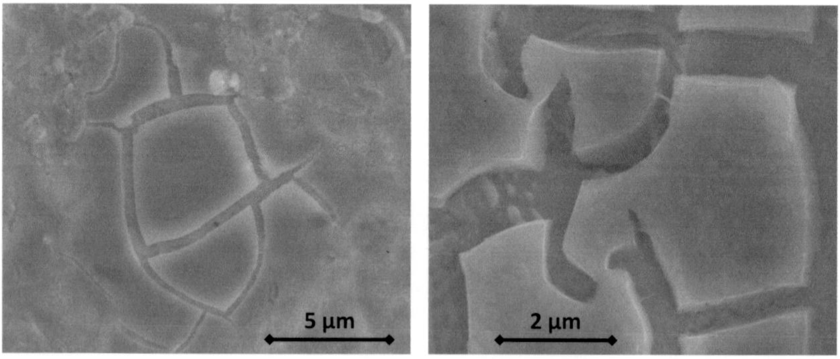

Figure 4.45. Crusty features on a fatigued as-cast sample (1.4×10^7 cycles, 200 MPa, $f = 1035\,\text{Hz}$, $R = -0.9$). The structure contains small particles as shown in the image (Right).

4.3. Discussion - Micro Molded Aluminum Bronze

The results, which were found in the investigations on micro molded aluminum bronze, are treated in this chapter. After analyzing the monotonic tests, namely tensile testing and nanoindentation, the fatigue tests, the microstructure and the damage morphology will be discussed.

4.3.1. Monotonic testing

The two tensile test curves (see figure 4.4) look very similar to the results on comparable samples presented by Rögner [55] (pp. 85-86). However, the stress parameters found here are higher and the strain at fracture as well as the Young's moduli are lower than the values found by Rögner. Even though the differences can reach several 10 %, the amount is comparable to differences Rögner found between different batches of samples. Tensile strengths for cast macroscopic samples made of CuAl10Fe5Ni5 are reported by Meigh [2] (p. 31) in the range of 551 MPa to 690 MPa. Thus, the strength values found for the micro samples (about 800 MPa to 850 MPa) are remarkably high. However, three other examples for die cast aluminum bronze given by Meigh (CuAl10.1Fe0.3Ni5.0Mn0.5, CuAl10.2Fe5.2Ni5.0Mn0.5, CuAl10.6Fe0.3Ni5.0Mn0.5) show tensile strengths of 765 MPa, 843 MPa, and 750 MPa with elongations of 9 %, 7 % and 5 % respectively [2] (p. 9). The results found in the tests of micro molded aluminum bronze in the present work are included in this range and could therefore be related to a specific chemical composition of the tested batch, given the tolerances in composition of the alloy (see section 4.1.1).

Nanoindentation imprints are visibly affected by the surrounding phases (see 4.2). The latter lead to the formation of pile-up and of slip bands around indents. This shows a mechanically discontinuous behavior of the material at the size scale comparable to the size of the precipitates.

The hardness values found with nanoindentation are in the range of $3.3 \times \sigma(\epsilon_{pl0.08})$ from the tensile tests, where $\sigma(\epsilon_{pl0.08})$ is the flow stress at 8 % plastic strain. This is in good agreement with general observations of Tabor for metals which work-harden, such as mild steels and annealed copper [88] (pp. 105-106), [89]. He showed for Vickers indentations that the (representative) deformation produced is about 8 % and that the Vickers hardness is about $3 \times \sigma(\epsilon_{pli} + 0.08)$. Herein, ϵ_{pli} is the initial plastic deformation of the material before indentation. These findings can be applied to the Berkovich indentations made on the aluminum bronze samples, since Berkovich and Vickers indentations induce the same representative strain [90] (pp. 26-27). However, a factor of 3.3 instead of 3 must be applied, when the hardness was measured with the Berkovich tip. (Remark: The explanation for the difference is the following: The Vickers hardness is calculated dividing the applied load by the surface area of the impression whereas the hardness gained with the Berkovich tip used in nanoindentation is calculated dividing the load by the projected area of the impression).

Summarizing the results from nanoindentation and from tensile testing it can be stated that the ductility of the aluminum bronze samples seems to be provided mainly from the matrix. The precipitates are harder than the matrix material and deform less.

4.3.2. Fatigue tests

Low cycle fatigue tests at frequencies of 100 Hz and 2.5 Hz as well as polishing the samples did not lead to remarkably different results (see 4.5). One might conclude a slight trend towards lower lifetimes at higher frequencies (for $R = 0.1$), comparing the two data points gained at 100 Hz and at 2.5 Hz with the data gained at 25 Hz. However, the lifetime does not change more than about 30 % per decade of frequency change.

As shown in the results, a difference in lifetime was observed between fatigue tests conducted at a load ratio of $R = 0.1$ and at roughly $R = -1$. The amount of differences in lifetime can be explained, applying the relations of Soderberg and Gerber [8] (pp. 131-132), which take into account the influence of a non zero mean stress applied during fatigue. The results indicate the presence of an endurance limit at about 190 MPa (no failure occurred below this amplitude at $R = -1$). This is at the lower end of the range reported by [2] (p. 31) for macroscopic samples made of CuAl10Fe5Ni5 aluminum bronze (190 MPa to 220 MPa). The related tensile strengths are reported with only 551 MPa to 690 MPa, which means that the strength values found for the micro samples (about 800 MPa) are remarkably high without increasing the fatigue strength in comparison to macroscopic samples.

The fatigue results are compared to the results found by Rögner in figure 4.46. Despite the different manufacturing parameters (mold temperature of 850 °C and 1000 °C) the results agree very well. It should be noted that the samples tested in this work as well as the samples tested in the work of Rögner were manufactured in different batches. Therefore, a certain variation in terms of overall microstructure can be expected, which should even be more pronounced when comparing the batches manufactured at different mold temperatures [54]. Thus, the overall microstructure does not affect the fatigue behavior remarkably. The difference in endurance found by Rögner (155 MPa) [55] (p. 91) and found in this work (190 MPa) can be explained again by the different R-ratios applying the relations of Soderberg and Gerber.

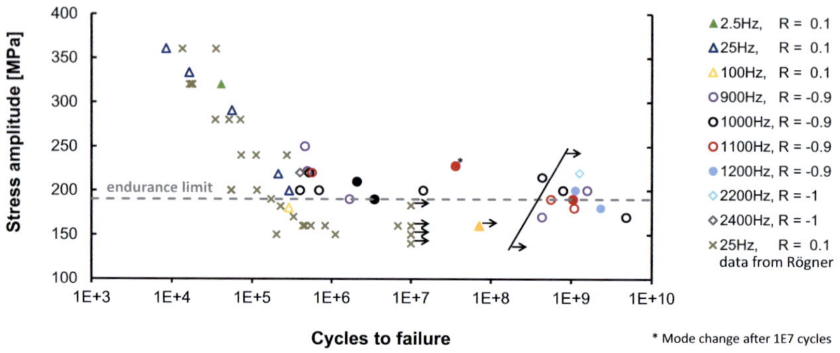

Figure 4.46. The data of Rögner (crosses) was added in the lifetime diagram of figure 4.5. Even though the batch investigated by Rögner was manufactured with a mold temperature of 850 °C, the lifetime data is very similar to the data found in the present work (mold temperature of 1000 °C).

4.3.3. Microstructure

The microstructure shown in the SEM images of the previous section is compared to the work of Hasan et. al. in figure 4.47. The morphology of the precipitates corresponds to the one reported by Hasan et al. [51]. However, their size tends to be smaller than reported in [51]. The two large dendritic-shaped precipitates in figure 4.47 (Left) look very similar to the κ_{I}-precipitate shown in 4.47 (Right), but they are more likely κ_{II}-precipitates for the following reasons: The κ_{I}-precipitate, shown in a micrograph from Hasan et. al. [51], has more dendrites than the dendritic precipitates found in our samples have, as indicated by the schematic derived from Hasan et al. of figure 4.47 (Right). Moreover, the κ_{I}-precipitate in the schematic is surrounded by α-matrix material, whereas the κ_{II}-precipitates are located in the vicinity of the lamellar κ_{III}-precipitates, which is also true for our microstructure. In addition, the difference in size between the κ_{I}-precipitate

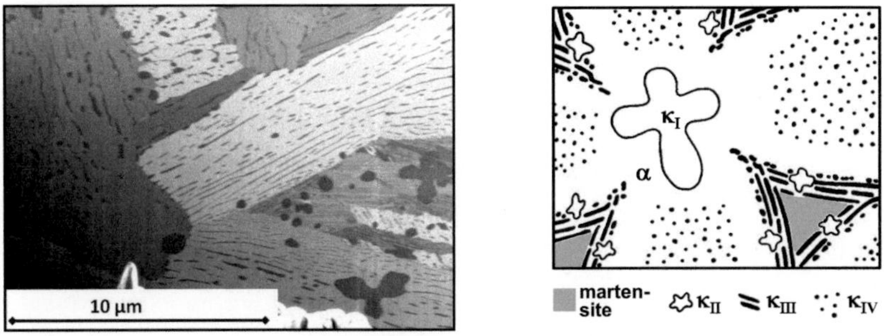

Figure 4.47. (Left) Microstructure of micro molded Aluminum bronze shown in a FIB image after milling with the FIB; (Right) Schematic derived from Hasan et al. [51] of the Aluminum bronze microstructure revealed by the authors.

and the surrounding microstructure from Hasan seems to be remarkably larger than the one observed for our samples. Hasan et. al. found κ_{I}-precipitates only in an alloy having a higher Fe-content than our alloy (Hasan et al. found them in an alloy with a Fe-content of 5.1 % but not in an alloy with a Fe-content of 4.4 %, [51]). Furthermore, the following sizes for the different kinds of precipitates are given in the paper of Hasan et. al. [51]: κ_{I}: 20 µm to 50 µm, κ_{II}: 5 µm to 10 µm. Thus, the more or less globular and the dendritic-shaped precipitates visible in 4.47 (Left) would be both suggested to be κ_{II}-precipitates, which have a size of 0.2 µm to 3 µm (herein, for the dentritic-shaped particles the rough circumference is taken). As discussed below, this holds true only for the larger dendritic-shaped precipitates, when EDX analyses are taken into account. The lamellar and degenerate lamellar (globular) precipitates are suggested to be κ_{III} precipitates. Their orientation changes from one matrix grain to the other. Thus, a preferred orientation relationship between the matrix and the lamellar precipitates is probable.

The smaller κ_{IV}-precipitates cannot be identified in this micrograph and are therefore expected to be smaller than 0.1 µm or not present in the visible region. The precipitates are not arranged in the same way as shown schematically in 4.47 (Right). The small grain size of the α-matrix seems to result in a more spread and less ordered arrangement of the precipitates. It can be observed quite often that precipitates are located at the grain boundary or at the free surface of the sample. The grain including the upper large dendritic-shaped precipitate contains very fine and very straight lines appearing dark and light. These lines indicate either an extraordinarily fine lamellar eutectoid of α and κ_{III}, or they represent the morphology of a fine lamellar martensitic retained β-phase. The latter appears as dark etching areas in the micrographs typically reported in the literature [51], [54].

The fine round-shaped precipitates from figure 4.7 (Right) show an arrangement and a morphology comparable to the κ_{IV}-precipitates identified by Hasan [51]. Also their size compared to the other precipitates indicates κ_{IV}-precipitates.

Areas containing fine parallel needle like stripes are clearly visible in figure 4.8 (Right), which appear dark and light. Bundles of them lying next to each other are oriented in different directions. This corresponds most likely to the martensitic retained β-phase. At some locations the needles seem to be more disordered and tend to overlap each other. One can also observe that the lamellar κ_{III}-precipitates contained in the light α-grain become smaller and more densely arranged the closer they are located to the grain boundary.

The attribution of the different precipitates to the phases is basically confirmed by the EDX analyses. According to Meigh [2] (p. 304), the precipitates with the highest Fe-content are the Fe_3Al-based precipitates κ_{I}, κ_{II} and κ_{IV}. Taking into account the relative precipitate sizes and the Fe-content of the investigated alloy of the present work, it can be concluded that the precipitates showing an elevated Fe-content in the EDX investigations are κ_{II}-precipitates. As discussed above, the κ_{I}-precipitates are most likely not present in our alloy and the κ_{IV}-precipitates are not large enough to be resolved with the applied EDX techniques. Thus, the large dendritic-shaped particles found in our samples are most likely always κ_{II}-precipitates, because they show an elevated Fe-content according to figures 4.9 and 4.10.

For the smaller round-shaped precipitates (about 2 µm or smaller in diameter), it seems more difficult to assign them to one precise type of precipitate. This is the outcome of the qualitative low voltage (5 kV) EDX measurements having a higher local resolution (see figure 4.11). The lamellar precipitates have a high content of aluminum and nickel, some copper and little iron. The same is valid for the round-shaped precipitates of about 0.5 µm to 1 µm in size. This agrees with [51] and [2] (p. 304) who describe the κ_{III}-precipitates as being lamellar or degenerate lamellar and based on NiAl. The Iron-rich precipitate in figure 4.11 is larger than the round-shaped NiAl-based precipitates, but both types cannot be clearly distinguished in terms of shape or SEM contrast. The elevated iron content corresponds well to the description of κ_{II}-precipitates by [51] and [2] (p. 304) as being Fe_3Al-based. One can conclude that most of the round-shaped precipitates are degenerate lamellar κ_{III}-precipitates, and some of them, in particular the larger ones (2 µm or larger in diameter), are κ_{II}-precipitates.

The high copper content found with EDX in the matrix material (see figure 4.10) agrees well with the description of the Cu-rich fcc α-phase (see Hasan and Meigh [51], [2] (p. 304)). An elevated Cu-content is shown in the same figure also for the regions with κ_{III}-precipitates, even though these precipitates are NiAl-based. This can be explained by the limited local resolution of the EDX investigations conducted at 20 kV. The matrix material encloses the precipitates. Therefore, matrix material in between and around the lamellar precipitates is also excited and adds to the signal. Thus, it is plausible that the differences in Cu-content as well as in Ni- and Al-content are visible in the analysis, but that they are at the same time less pronounced, than it could be expected looking at the chemical compositions reported by Meigh [2] (p. 304) and Hasan [51]) for the distinct phases.

The conclusion drawn from SEM-EDX results was confirmed by the higher resolving method STEM EDX, namely it is difficult to distinguish between round-shaped κ_{II}-precipitates and round-shaped κ_{III}-precipitates. The bright globular precipitate in figure 4.13 (about 800 nm diameter), which was analyzed by STEM-EDX (EDX spot S1), could be identified as iron rich (indicating κ_{II}). Other bright round-shaped precipitates of about 200 nm to 600 nm in size on this TEM lamella are rich in Ni, Al and Fe indicating κ_{III}.

In order to interpret the lamellar structures visible in the TEM analyses (see figure 4.12), mainly two possible explanations can be considered: either lamellar precipitates or lamellar martensitic structures. For the following reasons, a martensitic structure is the more convincing explanation:

STEM-EDX measurements on dark and light appearing lamellae show a high Cu as well as a low Al-, Fe- and Ni-content, similar to what would be expected for the matrix material (see figures 4.13 and 4.14). This agrees with the chemical analysis of the martensitic β phase conducted by Culpan and Rose and reviewed by Meigh [2] (p. 304). However, the EELS-measurements shown in figure 4.14 revealed a slightly elevated Ni-content along lamellae. This finding could be related to the very small dark appearing dots in figure 4.14 c), which are likely to be precipitates. This agrees with observations from Hasan et al. [51], who found NiAl-rich precipitates in the martensitic β-phase. They distinguished between two types of martensite: 3R martensite containing a high density of NiAl-based precipitates and 2H martensite. A presence of both types in our samples might be an explanation for the lamellar structures found by TEM analyses, which do sometimes show an elevated Ni-content (due to the contained high density of NiAl-based precipitates) and sometimes do not. Furthermore, the TEM images of the lamellae shown in figure 4.14 b)-c) remarkably resemble the TEM image of β-martensite from Sun et al. [91] figure 4 c).

Since the discussed TEM analyses were conducted on a TEM lamella extracted from the region shown in figure 4.47 (Left), the results can be correlated to the morphology observed in SEM images. Thus, the fine straight lines in the vicinity of the upper large dendritic-shaped precipitate from figure 4.47 (Left), which appear dark and light, can now be assigned to the β-martensite.

The small precipitates found in the lamellar structure of figure 4.15 are most likely the same type of precipitates in a lamellar martensitic structure as the ones shown in figure 4.14 c). The STEM-EDX signal presented in figure 4.15 (Right) (rising Ni- and Al-signals) is consistent with the finding from Hasan et al. [51] that NiAl-based precipitates are contained in the β-martensite.

In contrast, the small precipitates investigated in figure 4.16 are located in a lamellar NiAl-based precipitate (κ_{III}) and not in a region of lamellar martensite. This was shown by a STEM-EDX profile through the whole lamella with a similar result as shown in figure 4.28 (in addition, the signal in figure 4.16 next to the precipitate is comparable to the one found inside the lamella of figure 4.28). Furthermore, the morphology of the lamella observed in the SEM before extracting the TEM lamella (see figure 4.27) corresponded to the κ_{III}-morphology described and discussed above. EDX profiles of precipitates contained in the lamella revealed that they have an elevated copper content (see figure 4.16). The results must be interpreted with care, because the FIB extraction of the TEM lamella led to a slightly undulated surface (curtaining-effect) of the TEM lamella. E.g., the geometry of the elliptically appearing precipitates might be more circular in reality. Also a redeposition-effect of matrix material was considered to explain the elevated Cu-content. Yet, one would expect a more uniform deposition in this case, which is not restricted to the lamellar precipitates and which does not appear locally. Finally, it is more probable that there are Cu-precipitates in the investigated lamellar NiAl-based precipitate. In this context it is interesting to note that Hasan et al. found copper-rich precipitates inside κ_{I}-precipitates. The discussed results suggest that such precipitates can also arise inside the lamellar κ_{III}-precipitates.

4.3.4. Damage analysis

The various kinds of precipitates and elements contained in the micro cast aluminum bronze samples lead to various damage features, as it was presented in the results section. In the following, these features will be discussed and possible explanations for their formations will be given. However, some of these features in the results were mainly shown in order to give a catalog of findings for future analyses on such samples.

At first, features that were found on heterogeneous fracture surfaces will be discussed.

The linear structures with two preferential orientations shown in figure 4.19 (Left) are most likely oriented along crystallographic plains. However, they seem to be too large and too pronounced to be associated with slip steps. Therefore, it is suggested that these structures are related to fractured martensitic regions. The morphology fits well with the fact that different martensitic colonies are typically oriented along different directions. Some lines continue inside a cavity which formed most likely during casting (see upper left arrows). The hemispheric cavity in the lower left corner of the image has a diameter of about 6 µm. Inside this cavity, bumps are visible, similar to splashes. This observation suggests that the cavity was also formed during casting. The same is valid for the cavity shown in figure 4.20 (Left).

Assuming time dependent crack propagation (non catastrophic failure), the heterogeneous fracture surfaces shown in the results section indicate fracture along crystallographic orientations and an irregular crack propagation (different speeds - rest - continue - rest), typical for microcracks. A temporary crack propagation speed of about 0.12 µm per cycle is indicated by striations found on a small portion of the fracture surface in figure 4.19 (Right). The striations have a spacing of about 0.12 µm. The limited area showing these striations implies non-homogeneous crack propagation, depending on the local microstructure.

In a second step, features will be discussed, which indicate preferential crack propagation along interfaces between matrix material and precipitates.

Many examples were given in the results section, which show cracks propagating preferentially along lamellar precipitates, often oriented close to maximum shear directions (about 45° with respect to tensile load). It is most likely that slip occurring along the interface between matrix material and precipitates induces the observed cracks. First of all, indentations showed that the precipitates are harder than the matrix material, which favors slip activity in the matrix rather than fracture of the precipitate. Second, fractured round shaped precipitates were not found and lamellar precipitates were only fractured, when oriented at a large enough angle with respect to the crack (about 45° or more, see figure 4.23) - otherwise the crack tended to deflect along the direction of the lamella. Third, the cracks contoured round-shaped precipitates located in their path of propagation. Furthermore, the crack path in figure 4.25 is wavy like the surrounding lamellae, which is a hint for crack propagation along the interface between lamellae and matrix rather than for plain crack propagation inside the bulk material of the matrix oriented in parallel to the lamellae.

Taking these findings into account, smooth facets such as the one shown in figure 4.20 (Right) might come from fracture along a favorably oriented interface between a lamellar precipitate (κ_{III}) and the α-matrix. In contrast, slip steps or striations appear, as observed in figure 4.21, when the crack contours a globular precipitate and therefore changes permanently from one slip plane to another.

A further example for smooth regions on the fracture surface, which are assigned to fracture along the interface between favorably oriented lamellar precipitates and the matrix is shown in figure 4.34. The steps visible on the smooth surfaces can be related to kinking between lamellae arranged in parallel. This kind of kinking was observed in many examples such as figures 4.26, 4.35 (Right) and the crack faces in figure 4.38 (Right). One could also consider that these structures arise form the martensitic β-phase. However, the martensitic structures are expected to be finer and to change more often in direction, taking into account the observations concerning this phase discussed above. The granular structures on the fracture surface in figure 4.34 can be explained by abrasion due to crack closure. The sample was fatigued at an R-ratio close to -1 and the crack is oriented at about 45° with respect to the loading direction. Thus, pronounced crack closure can be expected during the compression part of a load cycle. Therefore, the smooth portions of the fracture surface might also arise partially from smoothening induced by frictional contact between the crack faces.

The main results indicating fatigue crack propagation along interfaces between matrix material and precipitates are summarized in figure 4.48.

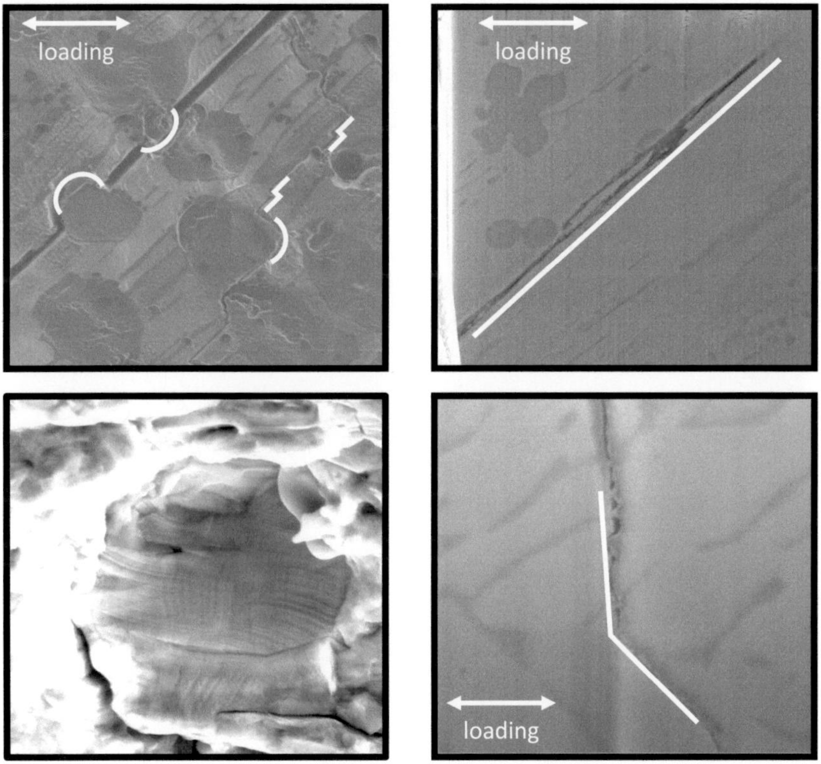

Figure 4.48. Summary of the fatigue crack behavior, which was observed in the HCF-regime up to 1.4×10^7 cycles: 1. Preferential crack propagation along lamellar κ_{III} precipitates, which are oriented at about 45° with respect to the loading direction. 2. Kinking of the crack towards crack guiding lamellae. 3. The crack is deflected by globular precipitates (round-shaped κ_{II} or degenerate lamellar κ_{III}). 4. Traces of slip activity are observed at the interface between matrix material and precipitates.

In a third step, extrusions found on cracks will be discussed in the context of oxidation.

A striking observation on cracked samples, which were not fatigued until separation into two parts, are extrusions coming out of the crack (they should not be confused with extrusions formed by slip steps). They can be curl-like, as shown in figures 4.22, 4.25 and 4.42, film-like as in figures 4.31 and 4.32, bubble-like (see figure 4.36) or bulge-like (see figure 4.41). The extrusions in figure 4.36 might appear bright because they have poor electrical contact to the sample surface and are thin, whereas the bulge in figure 4.41 is in better contact to the sample on a larger surface. Thus, the different kind of extrusions can have the same elemental composition and show different image contrasts because of their different geometries.

An obvious explanation for these features is that material is smashed out of the crack, after being abraded due to friction between the crack faces (pronounced crack closure at $R = -1$). However, the amount of material coming out, the sometimes pronounced localization and the height that these extrusions can attain is astonishing. Furthermore, these extrusions turned out to be mechanically well connected to the sample when they were extracted by means of a micro manipulator in order to investigate them using EDX (separated from the sample in order to prevent background signal).

Thus, oxidation was considered as an effect, which would increase the volume of the material and which would lead to a concentration of the extrusions at locations, where the material is rich of the oxidized element. Fatigue-induced damage would break the passivated sample surface, which would give access to the oxidized element and enable the oxidation which might even be promoted by fatigue. Fatigue-induced oxidation at room temperature was reported by Boyce et al. [66] for nickel, who explained this phenomenon by cracked natural oxide layers. According to the authors, the damage of the natural oxide is induced by persistent slip bands and gives the ambient oxygen access to non oxidized material. This is particularly interesting, because the cracks in the investigated aluminum bronze samples propagate preferentially along the NiAl-based κ_{III}-precipitates. Furthermore, it is interesting in this context that NiAl is known to form oxides at ambient temperature, which enhance electron transfer reactions compared to oxidized aluminum (see Lillard et al. [92]). In aluminum bronze, the eutectoide of $\alpha + \kappa_{\mathrm{III}}$ is preferentially attacked in corrosive media such as sea water or NaCl (see Wharton et al. [93] and Meigh [2] (p. 318)). In the review of Meigh [2] (pp. 318-325) it is stated that the martensitic β-phase is more susceptible to corrosion, and hence a maximum transformation of β into $\alpha + \kappa_{\mathrm{III}}$ during cooling or heat treatment enhances corrosion resistance. Furthermore, aluminum bronze is protected by an oxide layer against corrosion. These statements fit well with the observation made in figures 4.43 and 4.44. The sample surface is attacked in a region where the type of microstructure is located, which was assigned to the martensitic β-phase in the discussion above. Obviously, no corrosive medium was necessary to induce this attack.

Attempts were made to reveal the elements contained in the thin extrusions coming out of the crack using EDX. The results were difficult to interpret, because the thin extrusions induce a weak signal. However, they indicated an elemental distribution similar to the alloy, i. e. no preferential element seems to be involved in the formation of these extrusions.

In the future, it might be appropriate to conduct fatigue experiments in vacuum to get more insights in the role the oxides play during damage formation of cyclically loaded micro molded aluminum bronze.

The fourth step of discussing the damage analysis will be to explain the crack morphologies, which were found on the fatigued samples.

Typical traces of near threshold fatigue, which are well known for macroscopic samples (see section literature), were also observed on the investigated micro molded aluminum bronze samples. For instance, the corrugated crack paths in figures 4.30, 4.37 and 4.40 remarkably resemble the example shown by Blankenship et al. in their work on an Al-Cu-Li alloy (see Suresh [8] (pp. 209-210)), which Hornbogen and Zum Gahr attributed to glide along a single slip system (see Suresh [8] (pp. 209-210)). Furthermore, the work of Taylor and Knott on macroscopic aluminum bronze revealed the same kind of crack path morphology as the investigated micro samples of the present work show, but scaled up nearly one order of magnitude (see [20], [21]). The grain size of their aluminum bronze was about 100 µm (in contrast to barely more than 10 µm in the present work), and they found abnormally high and irregular crack growth rates for short cracks. Long crack behavior (moderate crack growth rates following the law of Paris) tended to appear when they reached a length of 300 µm (three grain sizes) [20]. Scaling down to the corresponding length in our samples one would obtain about 30 µm.

The central part of the crack in figure 4.32 (Center) is quite straight, whereas the upper part shows branching and the lower part shows film like extrusions. This might be an indication that the crack propagated easily in the center and was slowed down towards the extremities, due to a less favorable microstructure. Film like extrusions at the lower end as well as very localized chip like extrusions at the upper end and crack branching could arise from a more difficult and extremely slow crack extension.

As a result, many observations from macroscopic samples can apparently be scaled down to the micron range, since the manufacturing process leads to a corresponding downscaling of the microstructure. This suggestion was already made regarding the endurance limit; it is also confirmed investigating the crack morphology.

Sometimes, secondary cracks were found, which are oriented in one to three preferential directions (e.g. see figures 4.29 and 4.39). Several of these cracks are located apart from the main crack (largest crack or crack separating the sample into two parts). This indicates that multiple cracks are nucleated and propagate at different locations in the sample where the microstructure is oriented in a favorable way for crack nucleation.

In this paragraph, figure 4.39 will be discussed in detail. The conditions for crack nucleation induced by dislocation glide (high shear stress) are fulfilled at several locations depending on the local microstructure. Several cracks appear and indicate mainly three directions for mode II crack formation (sufficiently high local shear stresses) on the sample surface. The main crack in the center of the gage and the crack reaching the edge of the sample might have formed simultaneously and independently before getting as close together as they influenced each other. At this point they tended towards mode I propagation (perpendicular to loading). Secondary cracks are covered by dark bulges. Even extrusions come out of the crack, or formed cracks (eventually with extrusions coming out) are re-covered by an oxide. The main crack switches towards its tip between propagation along the direction of maximum shear (mode II) and the mode I propagation imposed by the mechanical far field load. This is visible in figure 4.40, where the loading direction is vertical. The crack propagates mainly at roughly 45° toward the loading direction (mode II) or close to the horizontal direction (mode I). The crack path is corrugated apart from straight parts in mode II direction. They might be flattened by pronounced friction along this direction due to crack closure.

Such behavior can be expected for near threshold fatigue, where crack nucleation takes the largest portion of the lifetime and compared to that, crack propagation takes only a rather small portion. The time of nucleation is large enough to induce cracks at several locations before one of these cracks starts growing continuously until failure. For fatigue clearly above the threshold amplitude, rapid localization and crack propagation of only the main crack would be more probable, because less time is available for crack nucleation at different sites.

The fifth step will be to discuss the damage behavior in the very high cycle regime and the impact of surface roughness.

For fatigue below the classical threshold amplitude, very high cycle fatigue behavior can arise. This means a failure mechanism appears after many load cycles (more than 1×10^8), which is for example fracture originating at inclusions in high strength materials (type II) [15]. In type I materials (fcc-metals, e.g. copper) such a mechanism would be surface roughening, which can induce micro-cracks after many cycles even below the classical threshold value. Roughening and microcracking typically appear in grains, which are favorably oriented for the formation of such damage, spread over the whole gage section of the sample (see [29]). A further possible reason for very high cycle fatigue damage reported in the literature is the elastic plastic incompatibility of phases in multiphase material, such as titanium alloys [15]. This phenomenon as well as the surface roughening could both be expected for the aluminum bronze, since the α-matrix is fcc and contains precipitates, which have other crystallographic structures and are harder.

However, only one from eleven samples failed above 1×10^8 cycles. Damage on the run outs could only be identified in few cases, mainly on the polished samples (the surface roughness of the as-cast samples is too pronounced to find such small features), and cannot be interpreted clearly. It seems probable that oxidation might play a role taking into account figures 4.29-4.32 and figure 4.42 discussed above. The damage morphology is similar to the one of samples, which attained lower lifetimes, but less pronounced. Only one very short and straight crack with curled features appearing at the crack tip was discovered on the samples shown in figure 4.42. It is believed that the stress amplitude was clearly below the fatigue limit in these cases. Hence, enough local plastic deformation, which forms visible damage, was only induced during cycling in the vicinity of stress concentrators such as the transition region of the gage section and the grip section (4.42 (Left)) as well as a casting defect (4.42 (Right)) combined with favorably oriented microstructural elements. In contrast, at higher amplitudes (i.e. at lifetimes of only several millions of cycles) visible damage could be formed at multiple locations (4.39) during the experiment, because at these load amplitudes the local stress is high enough to induce damage, even though the microstructural orientation is less favorable to induce slip activity. Thus, more locations with visible damage appear and the sample fails far before the VHCF regime is achieved.

Larger roughened areas as reported for pure copper by Stanzl-Tschegg et al. [29] were not identified on the aluminum bronze sample. One explanation might be that the lamellar precipitates within a grain of the fcc α-matrix lead to a localization of plastic deformation at the interface between both phases. Damage is formed at the location, where the combination of elastic plastic incompatibility between both phases and the orientation of maximum stress is the most critical. Areas without lamellar precipitates in this grain are subjected to lower local stresses. Regions of the damaged grain showing further lamellar precipitates might be subjected to lower local stresses because of a stress release caused by the damaged area or because of a less critical combination of elastic plastic incompatibility and orientation of maximum stress. Furthermore, the formation of glide steps inducing surface roughening might be hindered by an oxide on the sample surface. When this oxide is broken at a stress concentration, damage formation is localized there.

Another explanation could be that roughening appears mainly on the non-polished sample side-surface and cannot be discovered there. Of course it is not satisfying that the non-polished side-surface is a region of potential stress concentrators. When stress concentrators due to the as-cast side surface roughness would be the governing mechanism for damage formation, they could lead to failure of the sample before very high cycle fatigue damage, such as surface roughening, could appear. However, on the polished run outs discussed above, some damage formation was found on the polished sample (front- and back-) surfaces even though the sample did not fail. Thus, the roughness on the non-polished side-surfaces is not likely to induce remarkably higher stress concentrators. Furthermore, several cracks propagating at different locations from the side surface towards the center of the sample would be expected to appear on the polished sample surfaces, when stress concentrators on the side surfaces would initiate cracks. Since such behavior was not observed, and the fatigue limit is not lower than for entirely polished macro-samples, it is suggested that the roughness on the non-polished side surfaces is of minor importance. Possibly, the surface profile is smooth enough without showing sharp intrusions, or precipitates on the surface pores (as shown in figure 4.17) reinforce the local structure preventing crack nucleation. It might also be the case that stress concentrations induced by the surface roughness are less important than stress concentrations induced by the elastic plastic incompatibility of different phases, mainly of the very thin and long lamellar precipitates which could act like a stress concentrator comparable to a deep and sharp crack.

It is well known that the formation of dislocation arrangements such as dislocation cells is prior to surface roughening. Analyzing the consequences for ultra fine grained copper, Mughrabi concludes that a lower limit for the dislocation cell size formed in copper is 400 nm. Thus, for smaller grain sizes, the governing mechanism cannot be related to the arrangement of such dislocation cell structures. [9]

Since the spacing between lamellar precipitates in the aluminum bronze samples is sometimes smaller than 400 nm, it is likely that the formation of dislocation arrangements leading to surface roughening can be restricted by such microstructural features.

A set of parallel extrusions is shown in figure 4.38. They look like extrusions typically formed at locations of persistant slip bands due to surface roughening. Obviously, the stress concentration inducing pronounced slip is induced by the plastic zone of the crack tip. Probably, the lamellar precipitates, which were observed before to induce stress concentrations are not the dominating elements in this case. Therefore, the damage formation is not concentrated along one lamellar precipitate, but several parallel slip lines and extrusions appear in a matrix grain, where the stress is elevated in a larger area and not only along an interface. It is assumed that the lamellar stress concentrators are not present or oriented far from the direction of maximum shear in the observed area. Dark slip lines in the vicinity of slip extrusions and of cracks might be hints for slip steps, which are prevented to form properly by oxide layers.

In order to form such extrusions by surface roughening in the VHCF regime, a comparable zone size of elastic-plastically deforming material might be necessary to induce enough cyclic irreversibilities. However, due to the extreme localization of deformation to the interface of lamellar precipitates, the material volume which is mainly deformed might be so small that not enough statistic events related to irreversibilities can appear. Thus, no surface roughening due to slip irreversibilities is found.

4.3.5. Summary

In this discussion, it was shown that the ductility of the alloy is mainly provided by the matrix material, and the harder precipitates deform less. This knowledge helped to explain fracture morphologies and damage mechanisms. An endurance limit of about 190 MPa was determined, which is at the lower bound of data reported from macroscopic aluminum bronze. Thus, the fatigue properties are not affected by the small sample size, which can only be explained with a microstructure that is scaled down due to the micro molding process. Furthermore, the impact of the overall microstructure on the fatigue behavior is suggested to be small, since the fatigue experiments conducted at $R = 0.1$ in this work agree well with data found by Rögner on micro molded aluminum bronze manufactured with a lower mold temperature.

Compared to macroscopic aluminum bronze from the literature, the sizes of grains and precipitates is scaled down by roughly one order of magnitude. Furthermore, the precipitates are arranged in a more spread and less ordered way. Information from EDX, EELS and TEM helped to identify the different phases in SEM images. According to the discussion, κ_{I}-precipitates were not found in the investigated alloy. Most of the round-shaped precipitates are degenerate lamellar κ_{III}. However, they cannot be distinguished from κ_{II}-precipitates in the SEM images. Regions from FIB cuts observed in the SEM, which contain very fine and very straight lines were identified as β-martensite using information from TEM and EELS. Cu-rich precipitates were discovered inside some lamellar precipitates. Such precipitates are reported by Hasan et al. only inside κ_{I}-precipitates.

The origin of features which were found on the heterogeneous fracture surfaces could be assigned to fractured martensite, casting defects, precipitates and striations. Multiple indications were given for preferential crack propagation along the interface between precipitates and matrix material related to slip activity: SEM images of cracks running along or being deflected towards such interfaces, smooth regions on fracture surfaces which are assigned to interface cracking between lamellar precipitates and matrix material, slip traces in the matrix material at former locations of globular precipitates, preferential cracking along lamellar precipitates which are oriented in the direction of maximum applied shear stresses.

Extrusions coming out of the cracks were explained with material being smashed out of the crack due to the elevated frictional contact of the crack surfaces at an R-ratio of -1. However, it is suggested that oxidation could play a role in this context, because these extrusions are sometimes very localized, their volume is astonishingly high and they are mechanically well connected to the sample. The morphologies of the extrusions were curl-like, film-like, bubble-like or bulge-like. Since such observations are not reported in the literature, it would be interesting to investigate whether they are related to the material aluminum bronze or to the size scale at which investigations were conducted in the present work.

It was shown that the crack morphologies which were found for micro molded aluminum bronze in this work are very similar to observations reported for near threshold fatigue on macroscopic samples. However, the size scale is smaller in our case. Thus, the fatigue behavior of micro molded and macroscopic aluminum bronze is comparable, because the microstructure was scaled down with the micro molding manufacturing process, and therefore the fatigue crack formation is scaled down too. This is also in agreement with comparable endurance limits for macroscopic and micro molded aluminum bronze. A simultaneous formation of cracks at different locations was observed during near threshold fatigue without surface roughening. This supports the idea that multiple stress concentrations induced by lamellar precipitates at different locations cause fatigue damage formation.

The localization of deformation to the stress concentrations in the vicinity of lamellar precipitates is suggested as a major reason for the absence of a VHCF failure mechanism. This localization is due to the elastic-plastic incompatibility of precipitate and matrix material. However, it is the dominant failure mechanism already in the HCF-regime and therefore does not cause a new failure mechanism in the VHCF regime. Also, surface roughening, as it is reported for pure copper, is not found as a VHCF-failure inducing mechanism. Most likely, higher load amplitudes would be needed to generate surface roughening in matrix grains after about 1×10^9 cycles. However, such high amplitudes cannot be attained without inducing failure in the vicinity of stress concentrations caused by lamellar precipitates at lower cycle numbers in the HCF regime. Instead, one could expect surface roughening in the VHCF regime at even lower amplitudes in the vicinity of lamellar precipitates.

Surface roughening is related to random cyclic slip irreversibilities of dislocaions. Such a statistic process becomes less probable, the smaller the material volume subjected to slip activity is. Since in our case this volume is restricted to the vicinity of lamellar precipitates which are oriented in the direction of maximum applied shear stress, the number of random slip events can be tremendously reduced compared to a homogeneous polycrystalline material without stress concentrating precipitates. In the latter case, interacting random slip can be expected in volumes comparable to the size of a whole grain, assuming single grains which are favorably oriented for dislocation glide with respect to the applied load. The volume of random slip activity is not only suggested to be smaller in micro molded aluminum bronze due to the stress concentrating lamellar precipitates. Moreover, it might be more difficult to form dislocation structures which typically appear before surface roughening (cell or ladder structures) in this volume. The distance between lamellar precipitates was shown to be sometimes in the range of 400 nm for micro molded aluminum bronze. This is the order for dislocation cell sizes which appear in copper. Thus, the formation of characteristic dislocation structures leading to surface roughening might be sometimes prevented by small distances between precipitates.

5. Mechanical tests on micro molded Yttria Stabilized Zirconia

5.1. Experimental - Yttria Stabilized Zirconia

Three-point-bending tests were conducted on three differently manufactured batches of micro molded 3Y-TZP samples, in order to investigate their fatigue properties. Combining fatigue data with strength measurements, crack growth curves can be determined, which show the time dependent behavior of the material. Quasistatic bending tests were carried out to verify that the setup used in the present work produces strength data which is comparable to the data from the sample manufacturer. The latter was combined with the fatigue data in order to generate crack growth curves.

In addition, tensile tests on flat micro molded 3Y-TZP tensile samples were conducted to show the feasibility of such experiments on a small scale. The Young's modulus of the material was determined using image correlation from the tensile tests as well as from the three-point-bending tests.

5.1.1. Investigated samples

All 3Y-TZP samples investigated in this work were produced from a commercial ZrO_2-powder doped with 3 mol % Y_2O_3 (TZ-3YS-E, Tosoh, Japan).

Two different molding processes were used to produce the tensile geometry and the bending geometry respectively.

The tensile samples were manufactured by micro powder injection molding at the Institute for Applied Materials of the Karlsruhe Institute of Technology. The mold geometry was the same as for the production of the polymer model that is needed to produce the aluminum bronze samples (in the lost-wax lost-mold process). Thus, the flat tensile samples have the same dog bone shape as shown in figure 4.1 (Left) for aluminum bronze. However, the dimensions are reduced during sintering. Therefore, the resulting dimensions of the gage cross section are about $205 \times 115\ \mu m^2$.

The three-point-bending bars were manufactured by low pressure injection molding at the Institute for Applied Materials of the Karlsruhe Institute of Technology [94]. Their cross sections are squared (with rounded edges for one batch) having dimensions of $200 \times 200\ \mu m^2$. The length is about 1200 µm.

batch name	**Weibull strength [MPa]**	**Weibull modulus [-]**	**roughness R_a [µm]**
F4-3_A6_S6	1148	9.1	0.5
F2-2_A10_S10	1417	10.3	0.37
F11-4_A2_S2	2235	11.1	0.13

Table 5.1. Three different batches of 3Y-TZP bending bars were investigated in the present work. The table shows properties of these batches found with quasistatic three-point-bending tests by [95].

As described in [94], different feed stocks were used for different sample batches leading to distinct sample properties. For the present work, three batches were chosen to investigate the fatigue properties (table 5.1, [95]). Only the first part of each batch name (F2, F4 and F11) will be used to distinguish the three batches in this work. According to quasistatic three-point-bending tests performed by the sample manufacturer (table 5.1, [95]), the batches have different strengths. Thus, they can be classified into a "low strength batch" F4, an "intermediate strength batch" F2, and a "high strength batch" F11.

5.1.2. Micro tensile tests

The feasibility of tensile tests on micro molded ceramic samples was investigated using setup 1 (see figure 3.1). For this purpose, super glue was applied between sample and grips in order to prevent fracture during mounting. The Young's modulus was determined from the slope of the stress strain curves found by image correlation and load measurements. The load was converted into stress using the optically determined dimensions of each sample's cross section (images taken with the camera on the optical microscope from the setup, 14× and 35× magnification). The experiments were displacement controlled at a strain rate of $1 \times 10^{-5}\,\mathrm{s}^{-1}$.

5.1.3. Micro Three-point-bending tests

Three-point-bending tests were conducted with setup 1 adapted for bending tests. The setup configuration and the mounting procedure are described in section 3.1.4 and in [85]. The distance between the outer three-point-bending supports was 800 µm. Before testing, all samples were investigated with a confocal laser microscope (Keyence VK-9710k) in order to measure the sample dimensions and the surface roughness.

The widths of all four sample side surfaces were measured using camera images acquired with a 20× magnifying objective (Keyence VK-9710k). The roughness values were determined from the laser images on all four side surfaces, which were acquired with a 50× magnifying objective on the central part of the bending bar, where the stress is expected to be the highest during the three-point-bending tests. For this purpose, three 250 µm long line analyses were conducted in parallel to the sample edges, and the roughness was determined according to the Japanese standard JIS B0601 1994 (implemented in the Keyence software). Two of these lines are close to the sample edges, the third one is in the center of the sample surface.

Several displacement controlled quasistatic three-point-bending tests were conducted, the results of which could be compared to the data acquired by the sample manufacturer. Some of these tests were conducted at high displacement rates, leading to strain rates of about $5 \times 10^{-1}\,\mathrm{s}^{-1}$. This is comparable to the rates applied during fatigue testing.

A load ratio of $R = 0.2$ and a frequency of 100 Hz were chosen for the fatigue tests. Samples of the three batches (F4, F2 and F11) described in section 5.1.1 were fatigued, which show different inert strength levels due to different manufacturing parameters. Each batch was tested at only one load amplitude, because the available number of samples per batch was limited and the scatter of lifetime data is elevated. The load amplitudes were set to levels where few run outs and few spontaneous ruptures appeared leading to the following maximum stress values during fatigue: $\sigma_{maxF4} = 800\,\mathrm{MPa}$, $\sigma_{maxF2} = 1250\,\mathrm{MPa}$, $\sigma_{maxF11} = 2000\,\mathrm{MPa}$. A sample was categorized as spontaneously fractured, when it failed during the first few cycles before reaching at least 97 % of the target value in amplitude. Samples, which survived at least 1×10^6 cycles were considered to be run outs. They were fatigued at higher load levels afterwards until they fractured.

5.2. Results - Yttria Stabilized Zirconia

Before presenting the results of the different mechanical tests in this section, the roughness values measured on the bending cantilevers using the laser microscope are presented (see table 5.2).

batch name	roughness R_z [µm]	roughness R_a [µm]
F4	2.7±0.7	0.6±0.2
F2	0.7±0.3	0.3±0.2
F11	0.7±0.3	0.3±0.2

Table 5.2. Roughness values measured on the cantilevers from the three batches tested in bending. The roughness values are calculated according to the Japanese standard JIS B0601 1994.

5.2.1. Tensile characteristics

The stress-strain curves did not show a nonlinear behavior for higher loads, as it would be expected in the case of plastic deformation due to phase transformation. Spontaneous fracture occurred at 750 ± 20 MPa as shown in table 5.3. A Young's modulus of 231 ± 11 GPa was found.

sample	width [µm]	thickness [µm]	strength [MPa]	Young's modulus [GPa]
GM685-1	193	107	-	-
GM685-2	207	116	770	220
GM685-4	216	121	730	242
mean	206	115	750	231

Table 5.3. Dimensions, strengths and Young's moduli of the investigated 3Y-TZP tensile samples.

The brittle fracture of the samples is shown in figure 5.1. One sample fractured at the transition between gage section and clamping surface. The other sample fractured close to the middle of the gage section, whereas a slight lateral mismatch between the two sample fragments is visible.

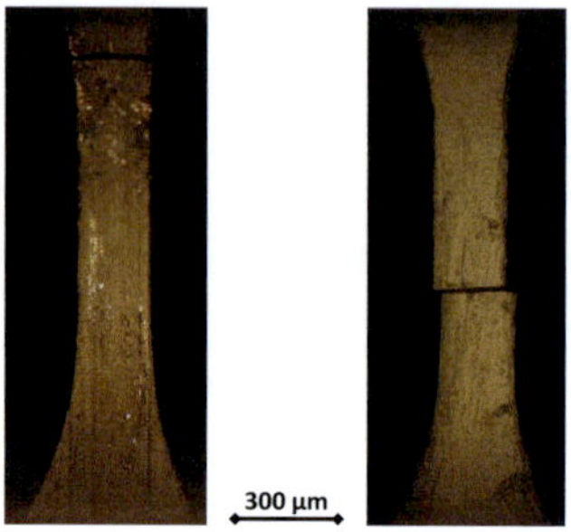

Figure 5.1. 3Y-ZrO_2 tensile samples after fracture: (Left) Sample GM685-2 (Right) sample GM685-4.

5.2.2. Strength values - three-point-bending

Stress-strain curves of three samples from batch F4 are shown in figure 5.2. A small hysteresis (barely detectable with the given resolution) is observed between loading and unloading segments for only one sample.

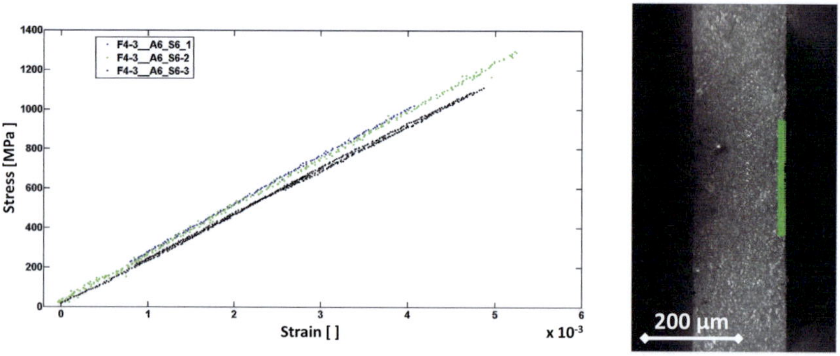

Figure 5.2. Three-point-bending: Stress-strain curves for the central part of the outer fiber deliver Young's modulus. On the right hand side the sample is represented with green markers on the outer fiber used for determination of the strain by image correlation.

The Young's moduli measured for each sample are given by the slope of the respective curve and are summarized with the strength values in table 5.4. The average Young's modulus is $237 \pm 10\,\text{GPa}$. Note, that the average strength of 1140 MPa concerns only the samples used for the determination of the Young's modulus, whereas further strength values were determined.

sample	width [µm]	thickness [µm]	strength [MPa]	Young's modulus [GPa]
F4-3_A6_S6-1	205	196	1020	243
F4-3_A6_S6-2	202	195	1290	241
F4-3_A6_S6-3	200	194	1111	227
mean	202	195	1140	237

Table 5.4. Dimensions, strengths and Young's moduli of the 3Y-TZP bending samples, which were used to determine the Young's modulus.

The strength values found with three-point-bending tests for the three different batches are shown in the diagrams of figure 5.3. Each data point represents one test. Even though the data sets are too small to generate reliable parameters of a distribution, the cumulative relative frequency is shown for a clear representation. The dashed lines indicate the Weibull parameter found by [95]. Since on average 63.2 % of the samples should fail at stresses up to this level, one can estimate roughly whether the data from the present work agrees with the respective Weibull parameter.

For the batches F2 (intermediate strength) and F11 (high strength), high speed tests were conducted at load rates (of hundreds of $GPa\,s^{-1}$) and strain rates (of about $5 \times 10^{-1}\,s^{-1}$), which are comparable to rates appearing during the 100 Hz fatigue experiments. The results are represented by solid symbols in the diagram. Fast tests lead to higher strengths than slow tests did. Inertia of the moving support sitting on the load cell was excluded due to experimental findings (fast motion of the system without sample) and analytic calculations. The solid lines in the diagrams mark the maximum stresses to which the samples were subjected during fatigue. Note that this level is remarkably high for batch F2 compared to the strength data.

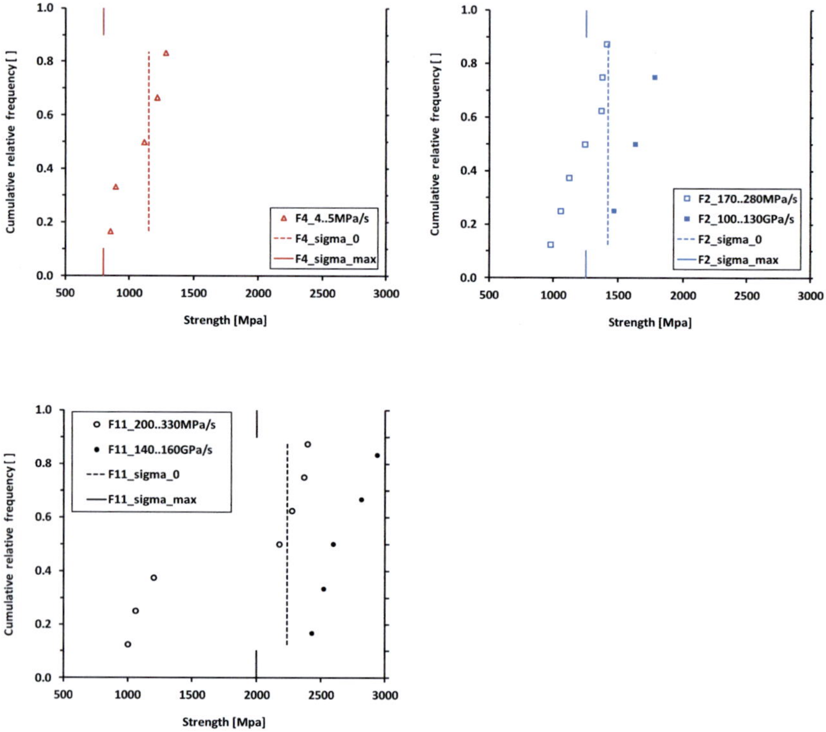

Figure 5.3. Cumulative relative frequency of strength values measured for the three batches of 3Y-ZrO_2 bending bars (one diagram per batch). Dashed lines indicate the respective Weibull parameter found by [95] (strength for 63.2 % cumulative failure probability). Solid lines indicate the maximum stress which was applied during fatigue tests for the respective batch. Higher load rates (solid symbols) lead to higher strength values.

5.2.3. Fatigue behavior - three-point-bending

The results of the fatigue experiments are represented in terms of Weibull plots in figure 5.4.

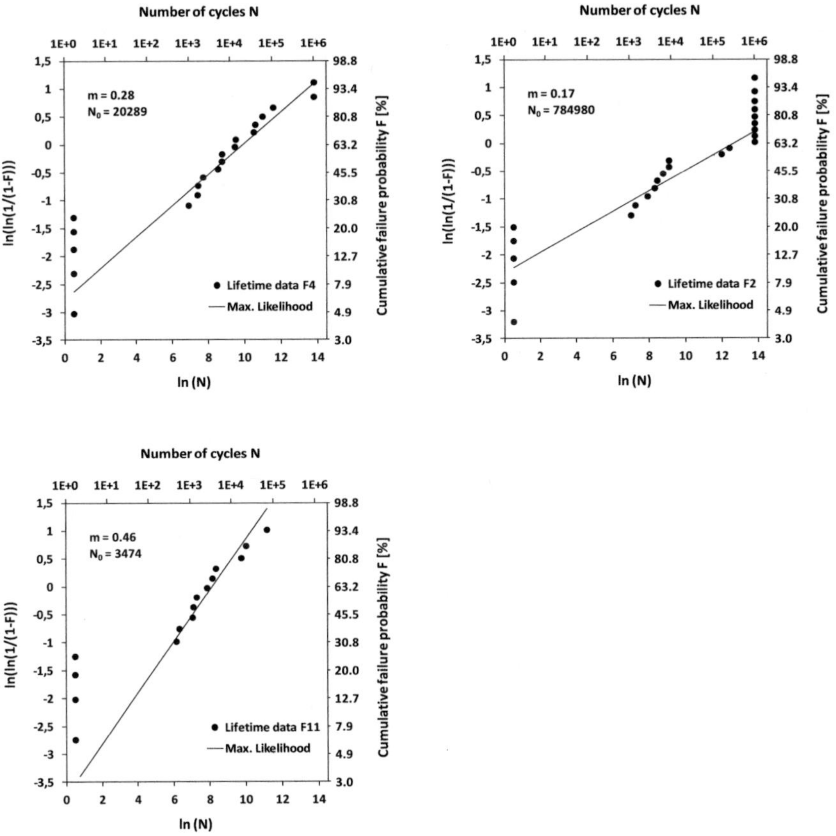

Figure 5.4. Weibull-plots of the three batches of samples (F4 - low strength, F2 - medium strength, F11 - high strength) which were tested in fatigue ($R = 0.2$; $\sigma_{maxF4} = 800\,\text{MPa}$, $\sigma_{maxF2} = 1250\,\text{MPa}$, $\sigma_{maxF11} = 2000\,\text{MPa}$). The solid lines show the corresponding Weibull-distribution found with the maximum-likelihood method, the parameters of which are written in the diagrams (m not corrected). For F2 and F11 more pronounced deviations from the Weibull-distribution are visible than for F4.

Spontaneously fractured samples are plotted with $ln(N) = 0.5$, i.e. $N = 1.6$, for better readability of the diagrams. Run outs are plotted with $N = 1 \times 10^6$, i.e. $ln(N) = 13.8$. The data of batch F4 follows a straight line in the diagram. Thus, it can be well described with the calculated Weibull distribution over the whole lifetime range. The Weibull-parameters determining the lines, which are plotted in the diagrams, were found by the maximum-likelihood method. Batch F2 shows a gap between 1×10^4 and 1×10^5 cycles where no failure occurred. For cycle numbers larger than 1×10^5 only two failures occurred, and a high number of run outs was observed. Batch F11 shows no run outs. Samples with low lifetime and samples with high lifetime are arranged below the Weibull line. The parameter m is a measure for the scatter (high m means low scatter). Accordingly, the scatter of lifetime data is the highest for batch $F2$, intermediate for batch $F4$ and the smallest for batch $F11$.

This lifetime data was combined with the strength data found by [95], in order to generate crack growth curves. For this purpose, the statistical method from [5] explained in section 2.3.2 was applied. The result is presented in figure 5.5. Calculations were performed using a geometry factor of $Y_\pi = 1.3$ (semi-elliptic surface flaw, see [5] (p.92) and [96] (pp. 7-25 to 7-33) and a fracture toughness of $K_{IC} = 4\,\text{MPam}^{1/2}$ for all batches. The fracture toughness corresponds to the maximum toughness values of 3Y-TZP reported by Rögner [55] (p.150) as well as to the plateau level of an R-curve found by Liu et al. [40]. Rögner conducted bending tests on micro molded 3Y-TZP samples containing notches of different lengths applied with a FIB. Liu et al. propagated cracks induced by indentation into macroscopic 3Y-TZP samples using four-point-bending.

The crack growth exponent of batch F4 $n_{F4} = 22$ is the lowest. Batch F2 has the highest crack growth exponent of $n_{F2} = 38$ and the high strength batch

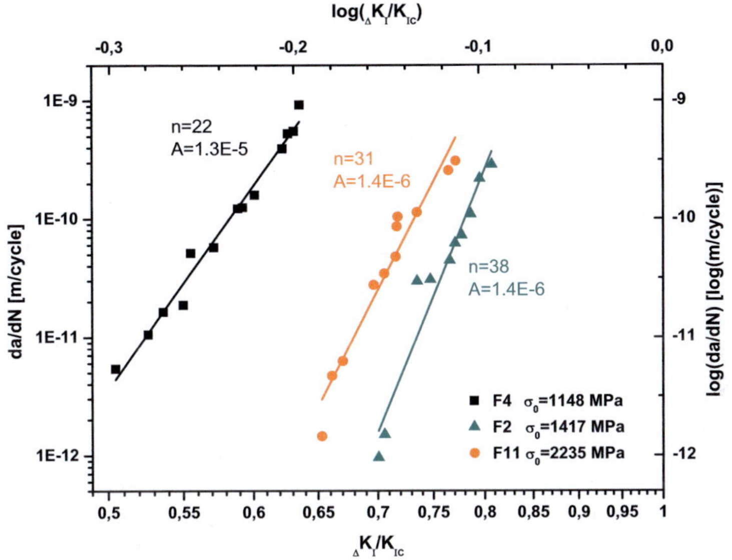

Figure 5.5. Three-point-bending: Crack growth curves found for three batches of samples combining strength data with fatigue data. The batches are named in the legend by the Weibull parameters σ_0 of the strength data. The curves were generated using $K_{IC} = 4\,\text{MPam}^{1/2}$ and $Y_\pi = 1.3$.

F11 shows an intermediate exponent of $n_{F2} = 31$. Assuming equivalent K_{IC} and Y_π values for all batches, the diagram illustrates that batch F2 was fatigued at the highest stress intensities with respect to K_{IC} (curve most far to the right). In contrast, F4 was fatigued at lower stress intensities with respect to K_{IC} and F11 lies in between the other batches. Note that spontaneous failures and run outs were included in the calculations but are not visible as data points in the diagram. Referring to figure 5.4, it becomes obvious that batch F2 has a pronounced number of run outs even though it was fatigued at stress intensities close to K_{IC}.

5.2.4. Microstructure and damage analysis

The grain structure of the three investigated 3Y-ZrO_2-batches is shown in figure 5.6, containing images of the as-cast surfaces. The images were acquired at 1 kV using the through-the-lens detector and give a mainly two-dimensional impression of the grains, which is favored to visualize the grain boundaries and to measure grain sizes. All batches show comparable grain sizes in the range of 200 nm to 600 nm.

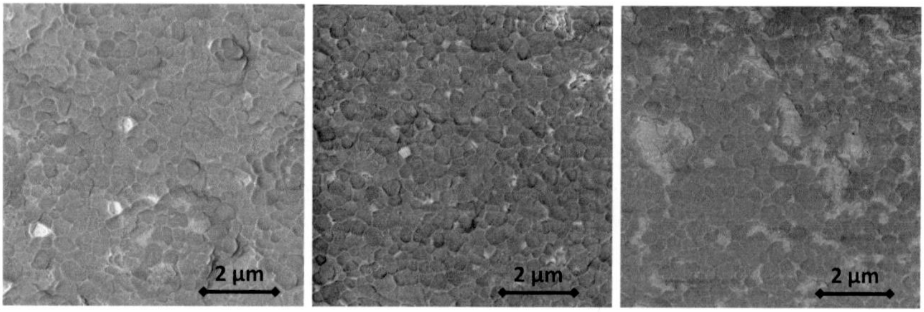

Figure 5.6. Images acquired with through lens detector: Comparable grain size for all batches varying in the range of 200 nm to 600 nm: (Left) Batch F2; (Center) batch F4; (Right) batch F11.

The three dimensional morphology of the grains on the as-cast sample surfaces can be seen in figure 5.7. The image on the left was acquired at 10 kV using the conventional Everhart-Thornley detector. A surface roughness on the order of the grain size (at the wave length of the grains) is observed, which is induced by the grains arranged on the sample surface. In addition, surface irregularities due to manufacturing (not in the image) contribute to the surface roughness. In the figure (Right), the grain structure is shown at higher magnification. For this purpose, a carbon layer (about 20 nm thick) was sputtered on a fractured sample of batch F2, in order to improve electrical conductivity of the sample. The pronounced surface profile related to the shape and the arrangement of the grains is visible. At the bottom of the image, the fracture surface (transverse to the image plane) shows some inter-crystalline and some transcrystalline portions close to the sample surface.

Figure 5.7. Three-dimensional grain structure on as-cast sample surface (samples from batch F2): (Left) Lower magnification (without carbon coating); (Right) Higher magnification after deposition of an approximately 20 nm thick carbon film to prevent charging, and hence to improve the image quality. The image shows the as-cast surface of a fractured sample ($\sigma_a = 500\,\text{MPa}$, $N = 6.4 \times 10^3$, $R = 0.2$). The fracture surface is transverse to the image and can be seen at the lower left corner of the image.

Regions with smooth facets as well as areas containing linear structures and hackles are visible on the fracture surface shown in figure 5.8. A sample from batch F11 was chosen. However, similar observations were made on fracture surfaces related to the other batches. The overview (Left) can help to estimate the overall portion of both surface types. The morphology of each type can be seen in the detailed view (Right). Both images are oblique views on the fracture surface (stage tilt of 52° corresponding to a view tilted about 38° away from the normal to the fracture surface). The same is valid for all following images of fracture surfaces if not described differently.

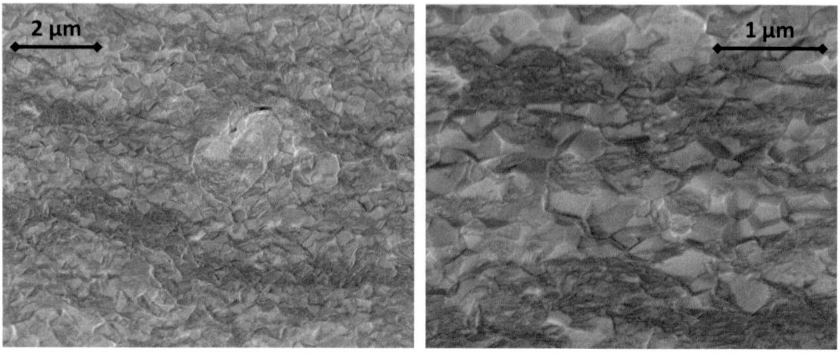

Figure 5.8. Fracture surface (batch F11 after 1.73×10^4 cycles at $\Delta\sigma = 800\,\text{MPa}$): Transcrystalline and intercrystallin portions of fracture.

Different flaw types were found investigating the fracture surfaces. Large pores, as shown in figure 5.9 (Left) were found in samples from batch F4.

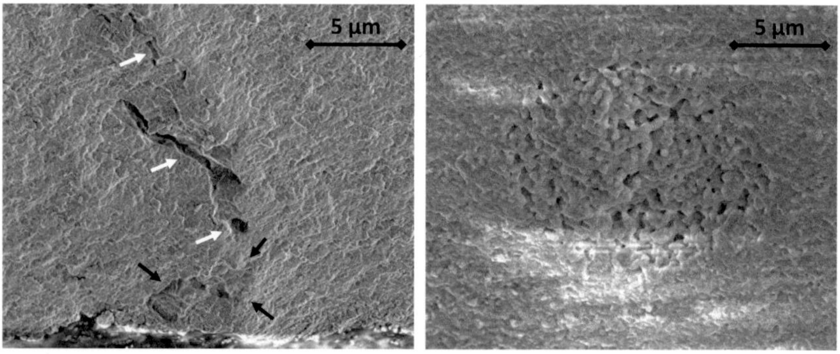

Figure 5.9. Flaws on fracture surfaces: (Left) Pores in a sample of batch F4 ($\sigma_a = 320\,\text{MPa}$, $N = 6.0 \times 10^4$, $R = 0.2$); (Right) Porous region in a sample of batch F2 ($\sigma_a = 500\,\text{MPa}$, $N = 1.6 \times 10^5$, $R = 0.2$).

The black arrows in the image indicate the circumference of a pore located at the sample surface (the sample surface is located at the lower edge of the fracture surface visible in this image). An elongated volume pore having a length of roughly 10 µm is located approximately 1 µm to 2 µm from the surface pore. It is extended by smaller neighboring pores towards the interior of the sample. These features are marked by white arrows. In the figure (Right) a porous region of 10 µm to 12 µm in diameter is shown, which was found on the fracture surface of a sample from batch F2.

A curled fracture surface of a sample from batch F2 is presented in figure 5.10 (Left). A large casting defect located at the corner of the fracture surface is indicated by a white arrow. This defect is magnified in the image (Right). The latter contains a view in the direction of the white arrow mentioned before.

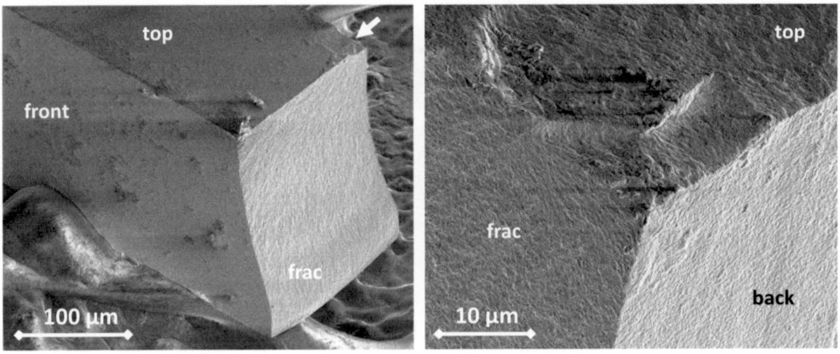

Figure 5.10. Large casting defect on the as-cast top surface, the location of which is slightly shifted towards the center of the bar, where maximum stresses occur. The crack plain is tilted towards the central loading support and curled (σ_a = 560 MPa, $N = 6.2 \times 10^2$, $R = 0.2$): (Left) Overview; (Right) Magnification of the casting defect.

A fractured sample from the same batch is presented in figure 5.11. Three flaws, i.e. locations of damage, were discovered on the top surface: A missing fragment at the upper left corner of the fracture surface, a surface pore surrounded by microcracks, and a surface pore at the upper right corner of the fracture surface. The latter extends to a step on the fracture surface which might be interpreted as a crack front.

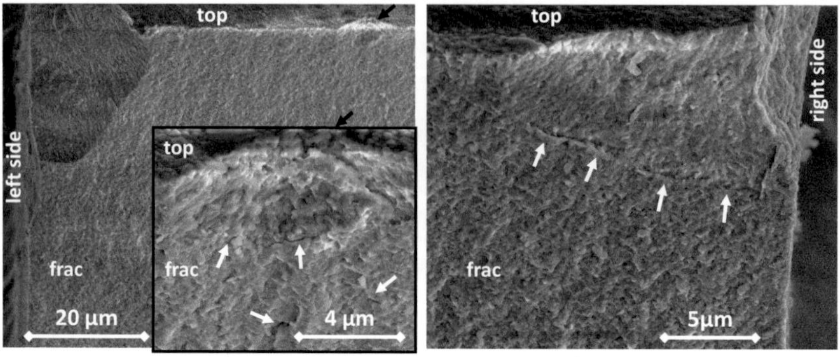

Figure 5.11. Damage at the edges of the bar and surface pores: (Left) Missing fragment at the upper left corner of the fracture surface and surface pore on the top surface indicated by a black arrow. This pore is magnified in the frame showing microcracks around the pore indicated by white arrows. (Right) Surface pore or missing fragment at the upper right corner of the fracture surface extending to the step indicated by white arrows. (σ_a = 583 MPa, $N = 1.4 \times 10^5$, $R = 0.2$ after run out at σ_a = 521 MPa, σ_a = 542 MPa and σ_a = 562 MPa).

A surface pore, which was observed on the fracture surface of a sample from batch F11 is shown in figure 5.12 (Left). Bright lines related to tiny surface steps can be seen, which run preferentially in radial direction with respect to the pore. The image in the figure (Right) presents damage on the fracture surface of a sample from the same batch. The sample edge (upper right corner of the fracture surface) is rounded, which is typical for all samples from this batch. A crack runs from the sample edge towards the interior in parallel to the sample surface. Thus, a fragment, which is not completely separated from the sample, was formed.

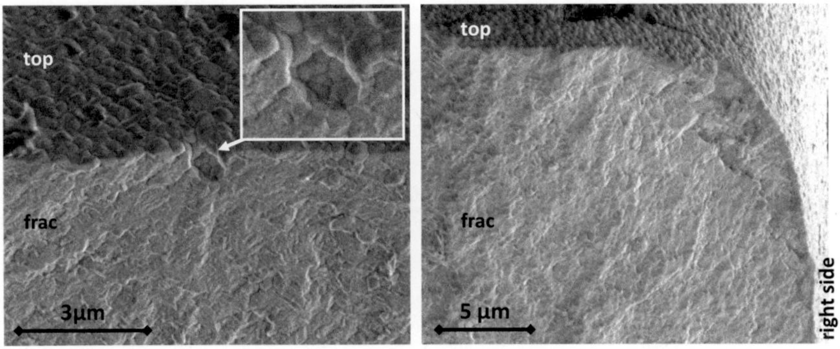

Figure 5.12. Fracture surfaces of two samples from batch F11: (Left) Surface-pore magnified in the detail ($\sigma_a = 800\,\text{MPa}$, $N = 1.7 \times 10^4$, $R = 0.2$); (Right) Rounded edge of the sample (batch F11 has rounded edges) with a microcrack running in parallel to the fracture surface from the right side surface into the sample forming a sliced fragment ($\sigma_a = 800\,\text{MPa}$, $N = 1.2 \times 10^3$, $R = 0.2$).

Further examples for pronounced steps formed on the fracture surface of samples from batch F11 are shown in figure 5.13. They are indicated by arrows.

Also traces of micro-cracks running into the fracture surface were found. Such a crack originating at the tip of a pore is shown in figure 5.14 (Left) (sample from batch F4). It is directed along a line connecting the sample edge with the central axis of the bending bar (diagonal of the cross section). The pore can be categorized as a subsurface flaw since it is located at about 5 µm from the sample surface and about 10 µm from the sample edge.

In the figure (Right), a fracture surface (batch F2) is visible, which contains a crack running from the sample surface into the interior. The crack propagates in parallel to the side surface of the bending bar (about 18 µm distant) to a depth of roughly 15 µm. The imaged location is at a crack depth of about 7 µm from the sample top surface. Crack branching and bridging are

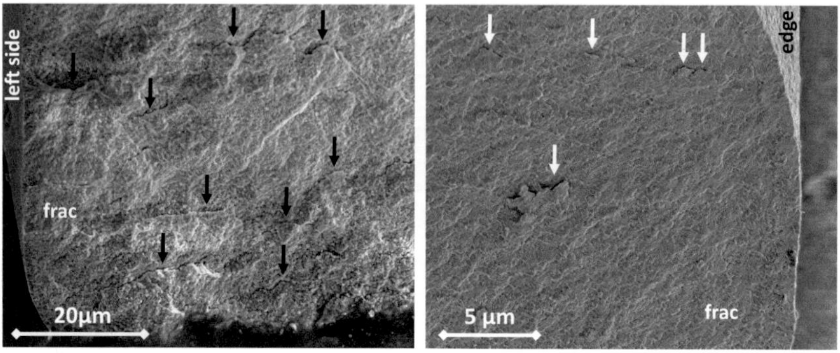

Figure 5.13. Steps on the fracture surface found in samples from batch F11: (Left) Microcracks running nearly in parallel to the fracture surface into the material form steps on the fracture surface, indicated by black arrows (monotonic test); (Right) A lower density of such steps was found on a fracture surface of another sample ($\sigma_a = 800\,\mathrm{MPa}$, $N = 1.2 \times 10^3$, $R = 0.2$).

observed. In both images of figure 5.14 linear structures are visible, which are oriented in parallel to regions having the size comparable to the grain size of the material.

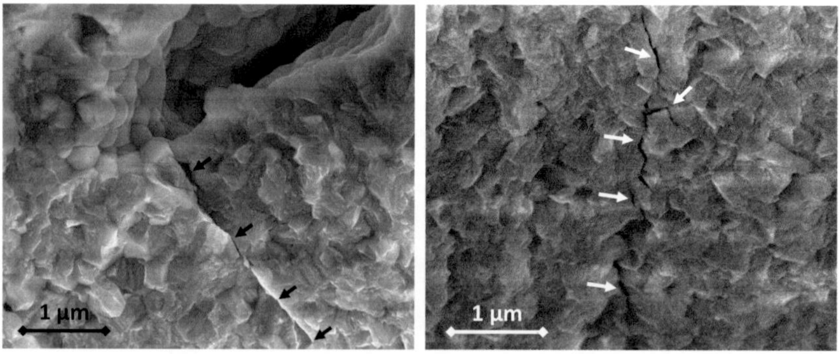

Figure 5.14. Fracture surfaces containing microcracks: (Left) Sample from batch F4 ($\sigma_a = 318\,\mathrm{MPa}$, $N = 1.7 \times 10^3$, $R = 0.2$): Micro-crack starting at a subsurface pore; (Right) Sample from batch F2 ($\sigma_a = 500\,\mathrm{MPa}$, $N = 1.4 \times 10^3$, $R = 0.2$): Micro-crack running in parallel to a side surface. The second white arrow from the top marks a location with crack branching, the second arrow from the bottom marks a location showing bridging.

5.3. Discussion - Yttria Stabilized Zirconia

In this section, the quasistatic tests will be compared to results reported in the literature and found by the manufacturer. Afterwards, the striking load rate dependence of three-point-bending fracture tests conducted in this work will be analyzed. It will be shown that subcritical crack growth has a major influence on strength tests conducted with micro samples.

In addition, the fatigue results will be analyzed. Flaw sizes and the amount of crack extensions until failure during fatigue will be estimated. The effect of the expected very small crack extensions on the development of crack tip shielding will be discussed from a microstructural point of few and related to the fatigue results.

In the next step it will be discussed which consequences can be expected for the R-curve behavior. The perceptions concerning shielding effects for small crack extensions are brought together with stress intensities applied during fatigue. As a result, it is suggested that the R-curve behavior starts only after a minimum crack extension length of about three grains. Consequently, an intrinsic toughness value of the material without shielding is related to fracture after smaller crack extensions due to subcritical crack growth.

In the last part of this section, the analysis of fractured samples in the SEM will be discussed.

5.3.1. Quasistatic tests compared to the literature

Strength data of micro molded 3Y-TZP tensile samples and bending samples was acquired and used to determine the Young's modulus of the material. Both testing methods lead to similar values of 231 GPa and 237 GPa. This confirms the feasibility of tensile tests with free standing ceramic samples having a dog bone shape, using the custom built setup (setup1). With a deviation of about 10 %, the value is not far from typical literature values of 210 GPa to 220 GPa for macroscopic zirconia samples (see section 2.5.1), especially when considering the typical scatter of micro sample tests and the low number of specimens tested here. By trend, the Young's modulus found in the three-point-bending experiments is elevated compared to the one found in tensile tests. One reason could be the small area that is available for strain measurements in three-point-bending, where stress gradients are present. Some of the markers used for image correlation are located in regions with stresses slightly lower than the calculated maximum stress in the bending bar. In addition, the curvature of the bar due to bending induces an axial marker displacement towards the central loading pin, which would lead to an underestimation of the strain, which would result in an over estimation of the Young's modulus. The stress strain plots in figure 5.2 contain unloading segments, which lie well on the loading segments and do not show remarkable hysteresis. Thus, no frictional problems occur using this bending setup.

A statistically meaningful set of three-point-bending strength data was collected by the sample manufacturer [95] for each batch, giving the respective Weibull strength and the Weibull modulus (see section 5.1.1). Data acquired with the custom built setup in the present work agrees well with these results, as it is demonstrated in figure 5.3. The open symbols represent strengths which were established at load rates comparable to the tests of the manufacturer and are mainly located close to the level of the Weibull strength (dashed lines in the figure). However, the set from the high strength batch F11 contained three tested samples with a remarkably low strength. The manufacturer found two comparable values out of 25 for this batch.

5.3.2. Rate dependence of quasistatic tests - subcritical crack growth

The three-point-bending tests conducted at elevated load rates (comparable to the rates of the 100 Hz fatigue tests) seem to show systematically higher strength values. Even though the limited number of data points is not sufficient for a vast analysis, it is interesting to discuss explanations for such a trend. For this purpose, a log-log plot of the strength versus load rate is plotted in figure 5.15.

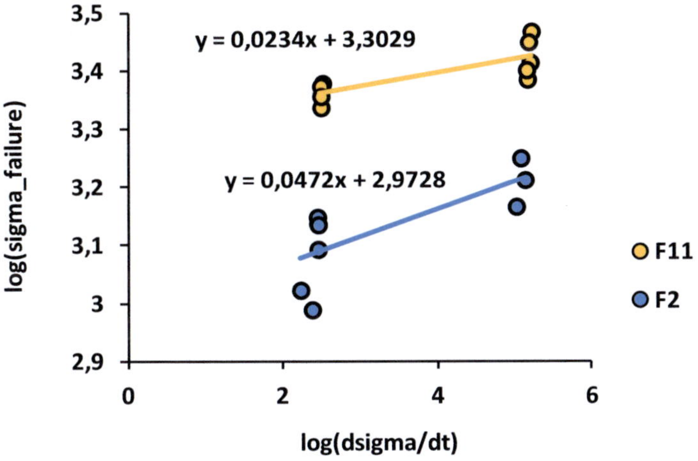

Figure 5.15. Rate dependence of the fracture strength in bending. The slopes of the linear fits can be used to calculate a crack growth exponent related to assumed subcritical crack growth. F11: high strength batch, F2: intermediate strength batch. The low strength batch F4 was not tested at elevated load rates, and therefore is not included in the diagram.

Assuming very fast occurring subcritical crack growth as the reason for a rate dependence, the related crack growth exponent can be calculated using the following equation [5] (p. 88):

$$n = \frac{1-s}{s} \tag{5.1}$$

Herein, s is the slope of the respective linear fit shown in figure 5.15 $s_{F11} = 0.023$, $s_{F2} = 0.0472$).

With this procedure, exponents of $n_{F2} = 20$ and $n_{F11} = 42$ are found for the respective batches. Three values of batch F11, which show an extraordinarily low strength, were not included in this analysis. The order of magnitude of these crack growth exponents is in the range of stage I subcritical crack growth (see section 2.3.1), taking into account the limited number of data points and the exponents found by [36] for the respective stages ($n_{\mathrm{I}} = 31.5$, $n_{\mathrm{II}} = 6.6$, $n_{\mathrm{III}} = 64$). Thus, a rate dependence seems to exist, which should be excluded for quasistatic three-point-bending tests applying high enough load rates or inert conditions. The unexpected high rate dependence might be due to a scaling effect concerning subcritical crack growth, i.e. stress induced corrosion.

Such a scaling effect might be explained as follows. In order to obtain the same K_I-value, higher applied stresses are necessary for small flaws than for large flaws (assuming the same conditions, such as location or shape for both types). Thus, the slope of the $K_I -_\Delta a$-curve for the smaller flaw is larger (see two curves with different applied stresses in figure 2.4 (Left)). Consequently, a small crack propagation due to subcritical crack growth leads to a larger increase of K_I (and therefore of the crack growth rate) for a small flaw than for a large flaw. Thus, a short propagation length due to subcritical crack growth, which occurs during the short time of the bending test might be negligible for a large flaw (in terms of rising K_I and rising crack growth rate), whereas, it is not negligible for the small flaw. Furthermore, the delivery of the corrosive medium to the crack tip might be facilitated for a small surface flaw, the crack tip of which is located closer to the sample surface.

The high speed fracture tests took about 0.02 s and the slower tests 7 s. In order to propagate the crack by about one grain (i.e. 0.4 µm) during one test, a mean crack growth velocity would be required, which is in the range of $0.4\,\mu m/0.02\,s = 2 \times 10^{-5}\,m\,s^{-1}$ and $6 \times 10^{-8}\,m\,s^{-1}$ respectively. These values correspond to the order of magnitude of the crack extension until failure estimated for batch F11 and F2 during fatigue, i.e. at stresses below the fracture strength (see table 5.5). They are in the range of $1 \times 10^{-8}\,m\,s^{-1}$ to $1 \times 10^{-4}\,m\,s^{-1}$ for which Chevalier et al. [36] found a power law according to the Paris regime. Consequently, for tests determining the strength at this small scale, very high load rates must be applied in order to avoid an impact of subcritical crack growth, allowing the experiment to be comparable to inert conditions.

5.3.3. Development of shielding during fatigue - crack extensions of only a few grains

Analyzing the fatigue tests, it is also helpful to investigate the crack propagation length which is necessary for failure. This is illustrated in table 5.5. The initial crack length (flaw size) is calculated from the Weibull-strengths (assuming $K_{IC} = 4\,MPam^{1/2}$ and $Y_\pi = 1.3$) with equation 2.1. The crack length inducing failure during fatigue is calculated with the same equation, in which now K_{IC} and the maximum stress σ_{max} applied during fatigue are inserted. Using both crack lengths, the necessary crack extension until failure is deduced. Thus, the crack extension for the high strength batch F11 is in the range of the grain size (400 nm). The flaws of the other batches extend by about three grains and nineteen grains respectively. Since the crack extension is on the order of a few grains, an interference between grain size and crack extension can be expected.

In this first approach, cyclic effects as well as any R-curve behavior are neglected. From experiments on macroscopic 3Y-TZP samples, a steeply rising R-curve is reported. If such an R-curve behavior also exists for small natural flaws in the micro samples, the fracture toughness of $K_{IC} = 4\,\mathrm{MPam}^{1/2}$ can be considered as the plateau value of the R-curve. The values for the above mentioned initial crack lengths calculated from fracture tests are then upper limits, because the R-curve behavior was neglected. This corresponds to fracture at K_{IC} without preceding crack prolongation. Due to overestimated initial crack lengths, the crack extensions until failure during fatigue might be larger than calculated. Experimental data showing an R-curve behavior of such small natural flaws is not available. In the following, the impact of small crack extensions on the development of shielding will be discussed. For this purpose, the calculated values will be used as estimates for the crack extension occurring during fatigue.

batch	σ_{max} **fatigue [MPa]**	**initial length [µm]**	**extension [µm]**	**# grains**
F4	800	7.2 / (-)	7.6 / (-)	19 / (-)
F2	1250	4.7 / (1.8)	1.3 / (0.5)	3 / (1)
F11	2000	1.9 / (0.7)	0.5 / (0.2)	1 / (0.5)

Table 5.5. Crack extensions until failure during fatigue assuming a fracture toughness of $K_{IC} = 4\,\mathrm{MPam}^{1/2}$ and ($K_{IC} = 2.5\,\mathrm{MPam}^{1/2}$) respectively. The initial crack length corresponds to the Weibull-strength (see table 5.1), the crack extension corresponds to a growth until K_{IC} is reached at the maximum stress applied during fatigue. The last column contains the crack extension divided by the grain size (400 nm).

Assuming a shielding effect, which is related to the grain size, different crack growth exponents of the three batches can be explained as follows. For instance, such a shielding mechanism could be friction and bridging between fractured grain boundaries. For batch F4, the crack extends through multiple grains. Thus, multiple bridges can develop along the crack path and the equilibrium between the formation and the degradation of shielding (bridging in our example), which is described in the literature (see [36], [43]) for fatigue of ceramics, can develop and determines the crack growth exponent of 22. This exponent agrees well with the exponents of 21 found by [40] and 20 found by [36] on macro samples.

In the case of batch F2, only a few grains (about three) are available for the formation of bridges. Therefore, it is difficult to develop an equilibrium of degradation and formation of the bridges. Even the crack propagates until failure without remarkable shielding effects, or it is heavily slowed, because shielding occurs, but degradation is very weak. For the degradation of bridges, interaction between crack surfaces is needed, which is limited for the crack having a length of one to three grains, since the interacting surfaces are very small for such a short crack. Thus, the probability of degradation is reduced. Furthermore, a minimum crack extension length is most likely necessary to form a bridge.

As is illustrated in figure 5.16 (Left), three grains can be considered being necessary to form a bridge. When a crack extends only about few grains, the major part of the extension (lifetime) is affected by the bridge, and it might not be completely degraded during lifetime. Therefore, one bridge should have a larger impact on a short crack than on a large crack. Also, the number of existing bridges has an impact on the importance of the formation of a single bridge. For short cracks, only about one or two bridges can be expected to be formed before failure. Thus, the impact is very high, whether a single bridge is formed or not. As a result, a tendency towards discrete behavior of shielding being relevant or not can be expected

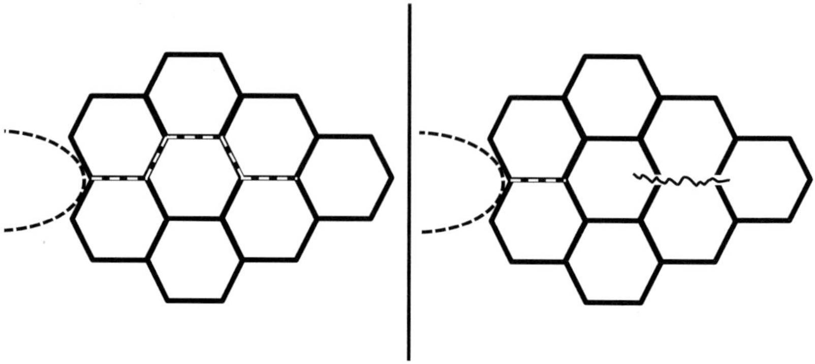

Figure 5.16. Model to explain cyclic degradation arising only after a minimum crack growth. Crack extensions having the size of about three grains are necessary to develop degradable shielding such as: (Left) Bridges; (Right) microcracks. The drawings show intercrystalline crack propagation. However, the same estimation can be made for transcrystalline crack propagation.

for cracks extending by few grains. This agrees well with the behavior observed for batch F2. Many run outs were found at the tested stress level (first fatigue test), which survived up to 1×10^6 cycles at remarkably higher stresses in additional subsequent fatigue tests. These might be specimens which were able to develop bridges, i.e. shielding leading to quasi crack arrest at the tested stress level. Samples which failed in the first fatigue test would then be specimens that did not develop shielding (such as bridging) due to their large initial flaw size, i.e. short crack extension until failure. The crack growth exponent of 38 for this batch is elevated due to the high number of run outs developing shielding.

According to this model, none of the samples of batch F11 develop degradable shielding, because the crack extension until failure is in the range of the grain size for all of them. As a result, the crack growth exponent is also high, namely 31. This is in good agreement with the exponent of 32, which was found in [36] for crack growth under constant loads (purely subcritical crack growth), i.e. for crack growth without degradation of shielding.

As discussed before, the high exponent 38 of batch F2 can be explained by two populations contained in the analysis of batch F2, namely samples with degradable shielding and samples without degradable shielding. Since the samples with shielding (run outs) do not fail during the fatigue test, the impact of the shielding degradation on the lifetime cannot be measured. Thus, all samples with shielding come into the analysis as if they would be run outs of samples without shielding. This high number of 9 run outs induces the high crack growth exponent.

In order to correct for this phenomenon, one can try to analyze only the population of samples without degradable shielding. The difficulty is in finding an objective criterion to assign the samples to the respective population. One approach could be to estimate the minimum crack extension to develop shielding. Let us assume that at least a crack propagation length of three grains is necessary to develop a complete bridge inducing degradable shielding, according to figure 5.16 (Left). The criterion for samples to develop shielding is then that their lifetime determining flaw grows at least 1.2 µm (three times the grain size) before failure. For this purpose, the strength data of batch F2 (acquired from [95]) was converted into flaw sizes applying equation 2.1 with $Y_\pi = 1.3$ and $K_{IC} = 4\,\mathrm{MPam}^{1/2}$. The flaws of the 14 strongest samples (46 % of the tested samples) would have to extend more than 1.2 µm. Thus, these samples were removed for a second crack growth analysis. Accordingly, the samples showing the 11 highest lifetimes (46 % of the samples tested in fatigue) were removed from the lifetime data. Using these reduced data sets corresponding to samples without shielding, a crack growth exponent of 28 was obtained. This value is close to 31, the value from batch F11. In this context, it is interesting to look at figure 5.4 (upper right graph of batch F2). The eleven samples showing the highest lifetimes are found after the absence of data that occurs between

1×10^4 and 1×10^5 cycles. The criterion, which was used to separate the two sample populations (crack extension of at least three times the grain size) agrees with the experimental observation that the 11 highest lifetimes do not follow the Weibull-distribution of the lower lifetimes.

In summary, the tested samples can be divided into three groups:

1. Samples showing purely subcritical crack growth during the fatigue test (batch F11 and first part of batch F2), because the crack propagation during lifetime is too small ($<$ three grains) to develop degradable shielding
2. Samples developing shielding at the end of their expected lifetime (lifetime if no degradable shielding would occur), which affects a high percentage of the relatively small crack length until failure ($\approx$ 3 grains). Therefore, the lifetime is remarkably higher than expected (second part of batch F2).
3. Samples developing degradable shielding. The crack extension is high enough to allow the formation of an equilibrium between degradation and formation of shielding ($\approx$ 20 grains). Thus, a crack growth exponent being smaller than the one related to subcritical crack growth is found (batch F4). This is the typical behavior for cyclic lifetime tests on 3Y-ZrO_2.

A discrete behavior is assumed for the whole model concerning the formation of degradable shielding. The fully developed shielding feature has a length of three grains. The idea was explained using the formation of bridges. However, it could also be applied for microcracking or phase transformation.

The mechanism, as it could occur induced by microcracking, is illustrated in figure 5.16 (Right) (see p. 198). The initial flaw must extend in order to form a propagating crack. Microcracks, which act as shielders in terms of stiffness reduction ahead of the crack are assumed to arise at a distance of at least one to two grains ahead of the crack tip (otherwise they would be part of the main crack). Furthermore, they must have a length of at least one grain to act as a microcrack, which explains the minimum length of three grains necessary to form a fully developed shielding feature by microcracks.

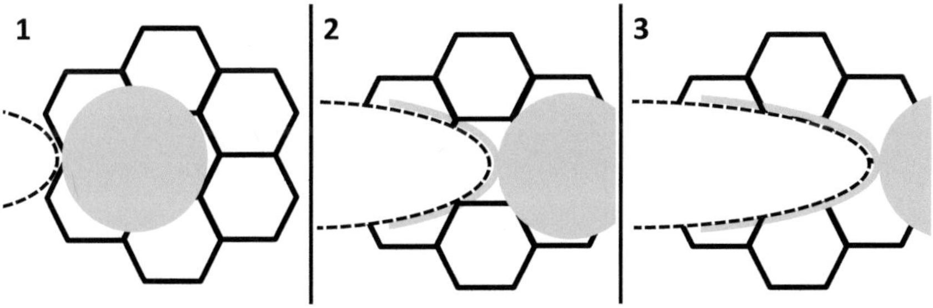

Figure 5.17. (Left) Formation of a transformed zone with a size of about one grain in front of the flaw; (Center) The crack propagates into the transformed zone and compressive stresses start to form around the extended crack line. (Right) The distance of the propagated crack, along which compressive stresses appear, is large enough to induce remarkable shielding.

Concerning shielding by phase transformation, it is assumed that the minimum volume which can be transformed is one grain. According to [5] (p. 63), the shielding effect caused by phase transformation is induced by compressive stresses arising along the wake of the extended crack. Dauskardt et al. analyzed load changes in transformation-toughened ceramics, and showed that shielding depends on the load history. After an increase in applied stress intensity K, a shielding effect occurs not before transformed material appears in the wake of the crack [97]. According to figure 5.17, it is plausible that a minimum propagation length of about three grains is necessary until the transformed zone, which is formed ahead of the crack reaches the wake and thus forms the degradable shielding.

An eventual influence of time-dependent phase transformation was also taken into account, analyzing the fatigue behavior of the three batches, and first of all the discrete behavior (run out or not) of batch F2. Time dependent phase transformation under uniaxial tension was found to induce creep at room temperature in Mg-PSZ by Finlayson et al. [46]. The authors explained the phenomenon by an anisotropic transformation strain in lenticular particles, which causes retarded transformation of less favorably oriented particles. One could imagine a similar effect for transformation of differently oriented and shaped tetragonal grains (having different sizes) in a polycrystal. However, in-situ X-ray diffraction [59] revealed phase transformation only very close to the fracture strength and assigned time-dependent anelastic behavior of 3Y-TZP at stresses below the fracture strength to grain reorientation or ferroelastic domain switching. These observations were shown by intensity changes of diffraction peaks during loading and by diffraction analyses of the fracture surfaces. The authors claim that this anelastic behavior is less pronounced at high load rates, i.e. high frequencies.

In figure 5.4, curve fitting of the data points, which are neither spontaneous ruptures nor run outs, would result in curves with negative curvatures. This could be expected for time-dependent behavior such as phase transformation in bending, where higher lifetimes lie below the expected straight Weibull-line [45]. According to the authors, the reason for such a behavior would be the reduced outer fiber stress due to a time dependent transformation. The discrete behavior of batch F2 in the present work could then be related to a spontaneous phase transformation and a related reduced outer fiber stress. However, the negative curvatures of the data fits are not pronounced and quite speculative. Assuming time-dependent phase transformation, it would be difficult to explain why there should be discrete behavior for batch F2, whereas a continuous phase transformation would occur for batch F4 (showing lower strength) as well as for batch F11 (showing higher strength than batch F2). Furthermore, 3Y-TZP is not expected to develop excessive phase transformation and the time-dependent phase transformation model was developed for more easily transforming materials, such as Mg-PSZ. Thus, time-dependent phase transformation is not likely to have a relevant impact on the presented experiments of the present work.

5.3.4. Expected R-curve resulting from the observed fatigue behavior and short crack extensions

Coming back to the hypothesis that a minimum crack extension is necessary to develop degradable shielding, one should think about the influence on the R-curve behavior and on the fracture toughness.

For this purpose, the basic idea is described in figure 5.18 (Left). Two curves show the stress intensities, which are applied at a flaw (having the intitial size a_i) during a strength test at the moment of fracture (orange curve) and during a fatigue test (green curve) over the crack extension $_{\Delta}a$. These curves are described by equation 2.1 containing the fracture strength (orange curve) and the maximum stress applied during fatigue (green curve), respectively. During a fracture test, the applied stress and therefore the stress intensity K_I rises continuously. Thus, K_I and the related crack extension $_{\Delta}a$ follow the initial R-curve until failure occurs at the tangent point of the orange curve to the initial R-curve. When a fatigue test is conducted instead, the flaw grows during the first fatigue cycle by a_1 (intersection of the green curve with the initial R-curve). During fatigue, the crack grows but the applied maximum stress per cycle stays at the same value. Thus, the applied maximum stress intensity follows the green curve during the fatigue experiment. The K_I-curve represents the applied stress intensity which is calculated using the flaw size and the applied stress. However, the respective stress intensity at the crack tip is expected to be smaller due to shielding. When crack tip shielding is degraded cyclically due to fatigue, the R-curve is affected, leading to lower R-values. This is indicated by the dashed curve, named degraded R-curve (see also schematic of reinforcement from Chevalier et al. [36]). Failure after fatigue is expected when the applied maximum stress intensity during fatigue (green curve) intersects with the degraded R-curve (crack extension a_2).

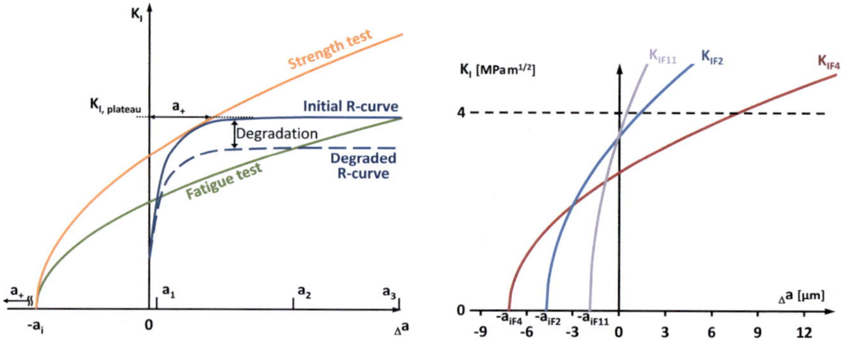

Figure 5.18. (Left) Stress intensities K_I over the crack extension $_\Delta a$ for a strength test at the moment of fracture (tangent point of the orange curve to the R-curve) and for a fatigue test. In the latter case, the initial flaw extends by a_1 due to load application in the first loading cycle. During subsequent crack extension, the applied stress intensity follows the green line until it intersects with the degraded R-curve (a_2), where failure occurs. When stress intensities are calculated neglecting the R-curve behavior, the K_I curves are shifted to the left by a_+, which is equal to the amount of overestimation of the initial flaw size; (Right) Stress intensities K_I over crack extensions $_\Delta a$ occurring during fatigue are calculated for the three investigated batches F2, F4 and F11, using the respective maximum stress applied during fatigue and the initial flaw size corresponding to the Weibull-strength from the fracture tests. The R-curve is neglected and a fracture toughness of $K_{IC} = 4\,\mathrm{MPam}^{1/2}$ is assumed.

As explained above, the initial crack length a_i is overestimated when the R-curve behavior is neglected (ideal brittle behavior), and the plateau value of the R-curve is inserted into equation 2.1 in order to calculate a_i. The amount of overestimation is visualized by the length a_+. The stress intensity curves in figure 5.18 (Left) would be shifted to the left by a_+. In this case, an ideal brittle material is assumed that fails in a strength test at the same stress and at the same stress intensity as the material with R-curve behavior. Thus, no crack propagation occurs before the sample fails at the stress intensity of $K_{I,Plateau}$. In the following, such shifted stress intensity curves will be used to discuss by trend the estimated stress intensities arising during fatigue of the tested 3Y-TZP batches.

The K_I-curves of applied stress intensities were calculated for the three batches using the particular initial flaw sizes a_i related to the Weibull strengths (see third column in table 5.5, p. 196), the maximum stresses applied during the fatigue experiments (see second column in table 5.5), and the fracture toughness $K_{IC} = 4\,\mathrm{MPam}^{1/2}$. According to the results (see figure 5.18 (Right)), the applied (maximum) stress intensity of the high strength batch F11 would be higher during the whole fatigue experiment (from $_\Delta a = 0$ to $K_{IF11} = 4\,\mathrm{MPam}^{1/2}$) than the stress intensities applied for the other two batches. Under these conditions it is not plausible to have a lower crack growth velocity for the small crack extensions of the high strength batch F11. However, the lower crack growth velocity would be needed to get lifetimes which are comparable to the results for the other batches. The same is valid, when the intermediate strength batch F2 is compared to the low strength batch F4.

Thus, the fracture toughness values of the batches F2 and F11 are most likely overestimated. Furthermore, it is suggested that no degradable shielding can be developed for the small crack extensions until failure of these batches (see discussion above: development of shielding during fatigue). However, the toughness must be lower without shielding (batches F2 and F11) than with shielding (batch F4). For these reasons, a fracture toughness of 2.5 MPam$^{1/2}$ is estimated for samples developing no shielding, leading to a smaller initial flaw size. This toughness value corresponds to the lowest level Rögner found in toughness measurements on micro molded 3Y-TZP [55] (p.150), as well as to the threshold level which can be estimated from the fatigue data on 3Y-TZP reported by Liu et al. [40]. As a result, the continuous curves K_{IF11} and K_{IF2} in figure 5.19 (Left) are found which help to explain the fatigue results. The dashed curves correspond to the stress intensity curves shown before, which were calculated with $K_{IC} = 4\,\text{MPam}^{1/2}$.

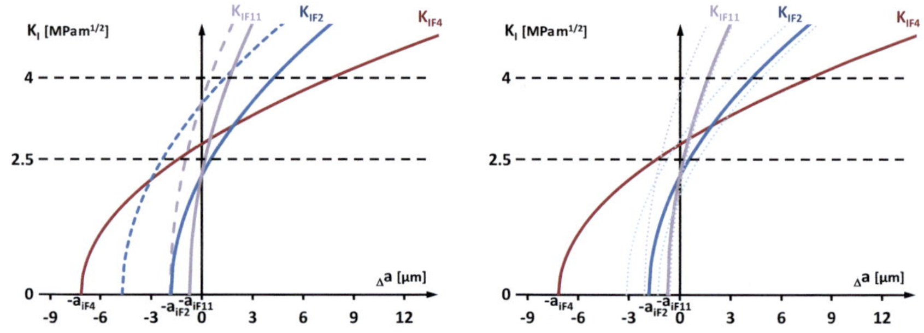

Figure 5.19. Modification of figure 5.18 (Right): (Left) A lower fracture toughness is assumed for batches F2 and F11 in order to explain similar lifetimes as found for batch $F4$. The solid lines correspond to the resulting new K_I curves. The dashed lines show the curves with higher toughness which were plotted in figure 5.18 (Right); (Right) Final stress intensities from the figure (Left): K_{IF4} is calculated with $K_{IC} = 4\,\text{MPam}^{1/2}$; K_{IF2} and K_{IF11} are calculated with $K_{IC} = 2.5\,\text{MPam}^{1/2}$. The solid lines correspond to the stress intensities applied during fatigue at flaws with initial flaw sizes related to the Weibull strengths. The dotted lines mark the stress intensities occurring during fatigue, which were calculated for the smallest and the largest initial flaw sizes corresponding to the largest and the smallest strength from the fracture tests conducted by the sample manufacturer.

Taking into account the distribution of initial flaw sizes, a series of K_I curves over $_\Delta a$ would describe the fatigue experiments of a whole batch.

Therefore, the K_I curves showing the highest K_I level (corresponding to the largest initial flaw size from the strength tests) and the lowest K_I level (corresponding to the smallest initial flaw size from the strength tests) are added in figure 5.19 (Right) (dotted curves) for batches F2 and F11. This configuration will be used to explain the following fatigue results (see figure 5.4):

- The high strength batch F11 showed three spontaneous fractures, a crack growth exponent of purely subcritical crack growth, and no run out.
- The intermediate batch F2 showed five spontaneous fractures, and a crack growth exponent of purely subcritical crack growth for 54 % of the samples; these having the smallest lifetimes (largest initial flaws). The rest of this batch showed abnormally high lifetimes 8.5 % and run outs 37.5 %.
- The low strength batch F4 showed a crack growth exponent which is typical for cyclic loads of macroscopic 3Y-ZRO samples, five spontaneous fractures, and two run outs.

Assuming an R-curve behavior during growth of the investigated natural flaws, one could explain fracture of samples from the batches F2 and F11 at small toughness values at the beginning of the R-curve. However, it seems difficult to give reasons for purely subcritical crack growth (i.e. no degradation of shielding) until failure for all samples of batch F11 without finding any run out. When no spontaneous fracture occurs, the stress intensity curve from fatigue must intersect with the R-curve at the beginning of the fatigue experiment, which means shielding starts to be established. In order to have purely subcritical crack growth without degradation of shielding and a lifetime which is not extraordinarily high, the stress intensity curve would have to intersect with the R-curve again after a very short crack extension. This would be necessary for all fatigued samples of batch F11 and for half of the batch F2. It could only be explained with different R-curves for different flaws showing a very smooth and precise shape over small crack extensions of only a few grains. Since shielding is related to the microstructure (see discussion above), such a scenario does not seem to be realistic. It is more likely that the R-curve behavior begins only after a minimum amount of crack extension, i.e., when the first shielding features start to be established. This idea is shown in figure 5.20.

At first, details will be explained considering the intermediate strength batch F2 (see curve K_{IF2}), which showed the extraordinarily high number of run outs. A flaw, having an initial size which corresponds to the Weibull strength, starts to grow during fatigue at the intersection of curve K_{IF2} with the vertical axis. In section *a*, it grows due to purely subcritical crack growth because no shielding is present which could be degraded causing additional crack growth. When the fracture toughness K_{IC}^{F11} is reached (the fracture toughness for samples of batch F2 without shielding is assumed to be equivalent to the toughness of batch F11), shielding starts to be developed because the stress intensity curve runs below the R-curve (section *b*). Due to this shielding, the crack is heavily slowed, causing a run out in the fatigue experiment.

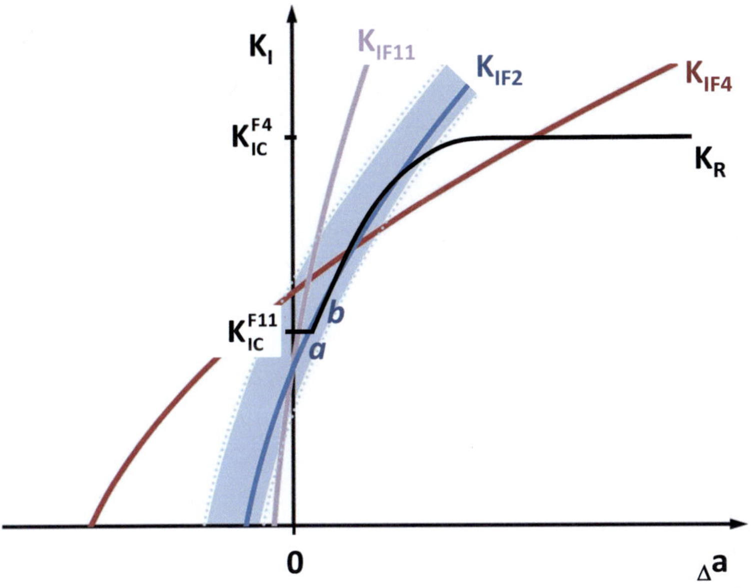

Figure 5.20. Schematic with R-curve: The applied stress intensities K_I from figure 5.19 are plotted over the crack extension during fatigue $_\Delta a$ together with a possible R-curve K_R. According to the previous plots, they indicate the maximum stress intensities occurring during the growth of a flaw due to fatigue, which has the initial size corresponding to the Weibull strength. The K_I-curves were again calculated using σ_{max} values and the initial crack lengths of table 5.5. They start at $K_I = 1\,\text{MPam}^{1/2}$ for a better visualization. Since the R-curve is neither calculated nor measured, the diagram axes are plotted without scale, and the graph represents a qualitative interpretation of the observed material behavior. The blue shaded area contains the array of K_I curves of batch $F2$, and is framed by the stress intensities corresponding to the minimum and maximum initial flaw sizes. These flaw sizes correspond to the highest and lowest strength values of batch $F2$ which were found by the sample manufacturer.

The blue shaded zone in the diagram represents the array of maximum stress intensity curves applied during fatigue for different flaw sizes of batch $F2$. It extends from the highest stress intensities on the left border, to the lowest stress intensities at the right border, which correspond to the largest initial flaw sizes and the smallest initial flaw sizes, respectively. This zone is divided by the R-curve.

For larger flaws, the stress intensity curve runs through the region below K_{IC}^{F11} (like section a of the blue curve K_{IF2}) and passes on the left from the R-curve, which corresponds to instable crack extension above K_{IC}^{F11}. These cases represent the lifetime data set without run outs (i.e. without shielding) and without spontaneous fracture. The lifetime is determined by purely subcritical crack growth (below K_{IC}^{F11}) before catastrophic failure occurs.

For even larger initial flaws of the blue shaded zone, the stress intensities at the beginning of the fatigue tests are higher than K_{IC}^{F11}, corresponding to spontaneous fractures. Small initial flaws with a stress intensity curve passing on the right of the R-curve are expected to induce run outs (as explained above for curve K_{IF2}). They are subjected to a fatigue experiment at stress intensities below K_{IC}^{F11} with purely subcritical crack growth and subsequently develop shielding at the end of the experiment.

According to the blue shaded zone in figure 5.20, arrays of stress intensity curves also exist in the vicinity of K_{IF11} (see dotted curves in figure 5.19 (Right)) and K_{IF4}, which correspond to the flaw sizes contained in the respective batches. They are omitted in the diagram for a better visualization. All these curves of batch F11 pass on the left of the R-curve, which fits with the observation that no run out was found for batch F11. No shielding is formed for samples of this batch, and all of them fail at K_{IC}^{F11} after crack extension related to subcritical crack growth, or they fail spontaneously during the first cycle because the stress intensity exceeds K_{IC}^{F11} from the beginning.

For most of the samples of batch F4, the stress intensity over $_{\Delta}a$ follows the R-curve due to application of load in the first cycle, until the maximum applied stress during fatigue is reached. Afterwards, the applied stress intensity rises due to growth of the fatigue crack and follows the stress intensity over $_{\Delta}a$ curve (e.g. curve K_{IF4}). The propagation is slowed by shielding and it is promoted by subcritical crack growth as well as by degradation of shielding. For a better visualization, a degraded R-curve (see figure 5.18 (Left)) is not added in figure 5.20. Also in batch F4, the largest flaws induce stress intensity curves, which do not intersect with the R-curve, and therefore cause spontaneous failure.

The reduced fracture toughness ($2.5\,\mathrm{MPam}^{1/2}$ instead of $4\,\mathrm{MPam}^{1/2}$) assumed for batches F2 and F11 does not change the crack growth exponents calculated for both batches. However, batch F2 was divided into groups of samples developing shielding and samples which do not develop degradable shielding, using calculations with $K_{IC} = 4\,\mathrm{MPam}^{1/2}$. The criterion for shielding development was a crack extension of three grains (fully developed shielding feature). When the same two groups of samples are formed and the crack extensions are calculated using $K_{IC} = 2.5\,\mathrm{MPam}^{1/2}$, the respective criterion for the development of shielding is only a crack extension of 1 grain. This would correspond to the beginning of the formation of a shielding feature (instead of the finalization of one feature). Therefore, the general idea that a minimum crack extension is necessary until degradable shielding can be formed, is not affected by the reduced fracture toughness.

In summary, two main conclusions can be drawn from the fatigue experiments in terms of R-curve behavior and toughness of the material:

1. When failure is induced by very small initial flaws (high strength batch) that grow by less than one grain (0.4 µm) until failure, the investigated 3Y-TZP ceramics show ideal brittle behavior, i.e., no R-curve behavior. In this regime, the fracture toughness is estimated to be in the range of 2.5 $MPam^{1/2}$ (constant fracture toughness before the R-curve starts to rise).
2. For larger flaws (low strength batch), an R-curve behavior is expected, leading to an increase of the fracture toughness up to about 4 $MPam^{1/2}$. In this case, a crack extension of more than one grain occurs in the first cycle of load application during fatigue and shielding is established. An equilibrium of shielding degradation and further formation of shielding is formed.

5.3.5. Fracture analyses using the SEM - small scale testing

Since we deal with micro samples failing due to very small flaws, it is mostly difficult to locate fracture origins and to conduct clear fractographic analyses. According to table 5.5, flaw sizes in the range of 0.7 µm to 7.2 µm (see third column) extending by only 0.2 µm to 7.6 µm (see fourth column) would have to be identified on the fracture surfaces. It seems that at this size scale, no classical mirror with surrounding hackle forms around the fracture origin, as it is known for macroscopic ceramic samples [96] (pp. 7-10 to 7-11), or it forms in a way that it cannot be clearly identified. One explanation could be that the mirror size which could be expected is on the order of the size of the fracture surface (and thus would cover nearly the whole fracture surface). For instance, macroscopic Y-TZP-samples fracturing at 1024 MPa can show a mirror diameter of about 100 µm [96] (pp. 5-15). Using the maximum stress values from table 5.5, an expected mirror size can be calculated using the following empirical equation [96] (pp. 7-10):

$$\sigma\sqrt{R_m} = A_0 \tag{5.2}$$

Herein, σ is the applied stress at the location of the flaw, R_m is the radius of the mirror and A_0 is a material constant, which is about 10 MPam$^{1/2}$ for 3Y-TZP [96] p. C-6). Thus, the following mirror diameters D_m would be expected for the tested batches: $D_{m,F4}$ = 313 µm, $D_{m,F2}$ = 128 µm and $D_{m,F11}$ = 50 µm. Only the mirror diameter of batch F11 would be small enough, with respect to the sample thickness of 200 µm, to develop in an ordinary way known from macroscopic samples. However, even this mirror size would exceed 30 % of the distance to the center of the bending bar, which is a recommended limit for fracture analyses based on mirrors [96] (p. D-25). Furthermore, mirror sizes in the range of only several 10 µm are on the order of other features such as hackles formed on the fracture surface. Therefore, it would be difficult to discover them. The formation of mirrors is affected by stress gradients [96] (pp. D-18 to D-19), and therefore could be suppressed in the investigated samples showing higher stress gradients than macroscopic samples due to their small thickness and high strength.

Sometimes it is challenging to collect the small fragments (not exceeding $200 \times 200 \times 600$ µm^3) of the broken samples and it is not evident to reconstruct their original position with respect to the supports. Light-optical microscopy provides a limited resolution and a limited depth of focus. Therefore, the fracture surfaces were investigated using the SEM. This implies difficulties due to charging effects and demands investigations at low voltages or after coating the sample with a conductive material. Therefore, fracture analysis of these micro samples is quite challenging.

Inter-crystalline as well as trans-crystalline fracture can be observed in figure 5.7 (Right) on the sample side surface, where it meets the fracture surface. Furthermore, inter-crystalline (smooth facets) and trans-crystalline portions (dark areas) of the fracture surface are visible in figure 5.8. The dark appearing structured surfaces contain partially parallel linear structures which are similar to fracture surfaces of ceramic single crystals (see [96] (e.g. pp. 8-18)), and are suggested to be related to cleavage along crystallographic planes (see [96] (pp. 5-60)). For the conducted experiments, it can be concluded that fracture occurs trans-crystalline or inter-crystalline, depending on the orientation of the grain or on the present phase. The latter aspect would be interesting to investigate. However, it would be necessary to establish Raman-analyses on the fracture surfaces of these samples with a resolution down to 1 µm. Another approach would be to conduct X-ray-diffraction. In order to reach the necessary resolution, it would be appropriate to use a synchrotron. However, an eventual tetragonal to monoclinic phase transformation occurring on the fracture surface would be difficult to show with this method, since it gives information not only from the material surface, but also from the interior volume.

Many observations of flaws and fracture traces are very similar to what can be found for macroscopic samples (scaled down to the micron size). Pores and porous regions shown in figure 5.9 are typically found in sintered ceramics. Large defects compared to the sample size, which would be introduced for example by machining in macroscopic samples, can appear in terms of casting defects in our samples, as shown in figure 5.10. Since the defect is not located symmetrically in between the supports inducing bending on the sample, the crack is twisted after initiation towards the stress concentration area in the center. As a result, a compression curl appears at the compression side of the sample, where the crack slows down and changes direction (see [96](page 4-17))

This behavior is also known from macroscopic samples. Likewise, missing fragments and surface pores, as shown in figures 5.11 and 5.12, correspond to observations known from macroscopic samples.

As mentioned before, the identification of mirrors on the fracture surface seems very difficult. However, structures similar to hackles, which would be expected in the vicinity of mirrors, were rarely found on the fracture surface and had relatively small characteristic dimensions.

Hackles originate from stress maxima of fast propagating cracks, which lie off the crack plane. Therefore the crack is locally twisted and small steps (hackles), oriented in the direction of the crack propagation, appear on the crack plain. [96](page 5-8)

Such structures could cause the bright appearing lines shown in figure 5.12, which point into the radial direction away from the surface pore (direction of crack propagation). More pronounced hackle like structures are shown in figure 5.13. They could also be interpreted as microcracks oriented at an angle to the fracture surface. Due to the small size of these structures compared to the fracture surface and its relief, it is not possible to use them as reliable indicators for the location of the fracture origin. For the same reasons, the fracture origin of the tested micro samples can rarely be determined accurately.

Branching and bridging, which are both important features inducing shielding, were found on a fracture surface, as shown in figure 5.14. Since the crack runs perpendicularly to the fracture surface, it did probably not induce remarkable shielding in the conducted experiment. It might arise, for example, from manufacturing (e.g. hot crack). However, the example shows that the features of discussion, which can induce shielding, are found in the 3Y-TZP micro samples.

The small sizes of the investigated samples have also an impact on the stress distribution versus microstructural sizes. The sample thickness of 200 µm and the elevated strength of the material induce a remarkable stress gradient in bending. Assuming a mean grain size on the order of 400 nm according to figure 5.6 and continuous isotropic conditions of a flawless bent cantilever, the stress decreases about ${}_{\Delta}\sigma_{0,F11} = 56\,\mathrm{MPa}$ per grain towards the neutral axis for the Weibull strength of batch F11 in the outer fiber (${}_{\Delta}\sigma_{0,F4} = 29\,\mathrm{MPa}$, ${}_{\Delta}\sigma_{0,F2} = 35\,\mathrm{MPa}$). This is equivalent of 2.5 % of the respective outer fiber strength. For macroscopic samples having the same grain size, much smaller stress gradients per grain are expected due to the following reasons. First, the cross section of a macroscopic bending bar, more grains are contained between the outer fiber and the neutral axes. Second, larger samples are expected to show a smaller Weibull strength due to scaling effects [5] (pp.145-148). Thus, the outer fiber strength is smaller, which leads to a smaller stress gradient.

The values in table 5.2 as well as the SEM images of figure 5.7 show roughness values in the sub-micron range, which are on the order of the grain size. These low values allowed mechanical testing of the as-cast samples without additional surface treatment. Thus, changes of the residual stresses due to surface machining could be avoided and the material was investigated as received from the manufacturing process.

6. Discussion - Mechanical properties at the microscale

Novel materials such as nanocrystalline nickel, nanotwinned copper, nanoporous gold, metamaterials, or micro molded materials, are often produced on a small scale. In order to investigate their mechanical properties, small scale testing techniques are required. The resulting need for custom built setups and the fulfillment of this need provided in the present work will be discussed in this chapter.

Furthermore, the major findings on the fatigue behavior of micro molded aluminum bronze and 3Y-TZP will be described. Micro molded aluminum bronze does not show a VHCF-failure mechanism, and 3Y-TZP shows cyclic degradation only when the crack extension until failure is large enough to induce shielding.

Afterwards, general implications of micro sample testing will be addressed. These result from the findings and the literature review of this work, as well as from current industrial requirements. The need for material tests on the microscale will be discussed from a material research point of view, as well as from an application oriented engineering point of view. It will be shown that the sample size, the microstructural size, and the testing equipment can strongly affect the experimental results of material tests at the microscale. Therefore, these three factors must all be understood and investigated. Thus, it is advantageous to combine research on experimental mechanics and material science.

6.1. Novel materials investigated due to custom built setups

In the present work, novel setups were designed which enlarge the pool of available testing methods for micro samples (having a width and a thickness on the order of 20 µm to 500 µm). Micro molded samples fall in this category, and were a large focus of this work. The flexible micro sample tester enables tensile tests, compression tests, as well as three-point-bending tests at quasistatic and at cyclic loading conditions with frequencies up to 100 Hz. These tests can be conducted with various kinds of micro samples. For fully reversed fatigue tests at higher frequencies up to 2 kHz, a resonant setup was developed. This set of small scale testing methods is needed to investigate and to develop novel materials which are produced or applied at dimensions of only hundreds of microns. Thus, material properties such as Young's modulus, Poisson's ratio, strength, yield stress, or the strain at failure can be determined, and the reliability in terms of lifetime under cyclic loading can be investigated for samples on the microscale. For instance, the setups designed in this work were used for the following experiments: compression tests of ceramic pebbles having diameters of 500 µm [83], fracture toughness tests of nanocrystalline nickel-tungsten coatings [85], fatigue tests of nanotwinned copper [72], tension and compression tests of nanoporous gold [81], compression tests of polymer metamaterials [84], micro molded aluminum bronze (this work and [98]), and micro molded 3Y-TZP (this work and [99]).

The behavior of the test rigs was optimized for micro sample testing in terms of speed, control parameters, and reliability for long term tests. In addition to the proper hardware selection, this was achieved by programming the whole software for data acquisition and control. Software development was an important part of the present work, and was realized with Labview. The main components of this software were also used for the control of further setups (for tensile testing, fatigue, and multiaxial fatigue, see [85, 100]), which were designed in the small scale reliability group of the KIT. In addition to the software, the compact design of the setups and the selected components provide a highly dynamic behavior, due to their high stiffness and their elevated accuracy. For instance, the piezo actuator is very stiff and the low weight of moving parts such as grips or load cells cause only small inertial loads. Furthermore, the Field Programmable Gate Array data board provides non stop real time control of the piezo motion with a resolution in the nm regime. Such properties enable a more precise loading of micro samples than can be realized with commercial heavy testing machines, which are typically designed for macroscopic samples.

As a result, it was possible to investigate samples at the end of their lifetime, which contained fatigue cracks having a length of less than 100 µm, but were not yet separated into two parts. Therefore, the interaction between fatigue damage and microstructure could be investigated for micro molded aluminum bronze. Furthermore, fatigue experiments could be properly conducted at 100 Hz with brittle zirconia bending bars having cross sections of $200 \times 200\ \mu m^2$.

With the resonant setup, micro molded aluminum bronze could be fatigued in tension-compression at frequencies up to the kHz regime. Thus, even the VHCF-regime could be investigated in a reasonable amount of time. Since the Eigenmode of the sample is excited, load ratios close to -1 can be realized, even though the samples are flat (cross sections of $260 \times 130\ \mu m^2$). In contrast to conventional VHCF-investigations, experiments on the novel resonant setup are load controlled. Continuous cycling is possible without creating problems of self heating, because the experiments are conducted at a frequency of about 1 kHz instead of the common ultrasonic testing frequency of 20 kHz. Furthermore, the elevated surface to volume ratio of micro samples is advantageous to dissipate heat. Thus, a new frequency range is now available to conduct fully reversed tension-compression HCF and VHCF-experiments at the microscale.

Due to the sensitivity of the setup for fatigue damage, the experiments could be stopped a few cycles before final failure. Therefore, cracking along lamellar precipitates was identified as the predominant failure mechanism of micro molded aluminum bronze in the HCF regime and globular precipitates were found to deflect fatigue cracks. However, no VHCF failure mechanism was found.

Two main reasons are known for fatigue failure in the VHCF-regime. First, failure can originate at inclusions (e.g. in high strength steels), leading to a multistage lifetime diagram. Second, surface roughening due to slip irreversibility can induce fatigue cracks which propagate until failure. None of these mechanisms were found for micro molded aluminum bronze, which would induce failure at low loading amplitudes between 1×10^7 and 1×10^9 cycles.

Apparently, internal VHCF-failures originating from inclusions or phases at low amplitudes are specific for only few materials such as high strength steels or titanium alloys. Major factors in these cases might be the missing ability of matrix material to accommodate for deformation in the vicinity of less deforming inclusions or phases as well as the involved interfaces. Since this appears to be critical for the investigated aluminum bronze already in the HCF-regime (fatigue cracks along lamellar precipitates), it cannot arise as a new failure mechanism in the VHCF-regime.

The second well known VHCF-failure mechanism, namely surface roughening as it is found for pure copper, can obviously be prevented in aluminum bronze by adding alloying elements, whereby dislocation motion is hindered. An enhanced formation of surface oxides might also prohibit the formation of extrusions and subsequent surface roughening. As a result, surface roughening in the matrix would arise after 1×10^9 cycles only at rather high stress levels. However, these stress levels cannot be attained without inducing failure at lower cycle numbers due to locally elevated shear stresses at unfavorably oriented lamellar precipitates. Thus, an endurance limit was found for micro molded aluminum bronze fatigued up to 1×10^9 cycles. Its level is most likely defined by shear stress concentrations induced by unfavorably oriented lamellar κ_{III}-precipitates. Fatigue failure of the investigated as-cast micro molded aluminum bronze can therefore be prevented even in the VHCF-regime, when the endurance limit is not exceeded. Consequently, the alloy can be recommended for applications in terms of reliability of micro molded components.

A further interesting candidate of micro molded materials is zirconia. Three-point-bending tests of micro molded 3Y-TZP samples were performed with the flexible micro sample tester. The reliability of such micro samples can now be better estimated taking into account the results of the present work. It was shown that the high strength batch subjected to cyclic loads has a crack growth exponent which is reported by Chevalier et al. for purely sub-critical crack growth of 3Y-TZP macro samples under constant loads (see [36]). This could be explained with natural flaws, which extend during fatigue only about one grain until failure. It was concluded that degradable shielding cannot be developed during such a small crack extension. Therefore, no cyclic effect in terms of shielding degradation can occur during fatigue. However, the low strength batch shows such an effect, because it contains larger flaws, and the samples are subjected to lower stresses during fatigue. Thus, the increase in stress intensity due to crack propagation is smaller and the crack extension until failure is elevated. The latter is high enough for shielding to be formed, which can be degraded during cycling. This could explain the crack exponent which was found for this batch, at a level that is reported by Chevalier et al. for cyclic degradation of 3Y-TZP macro samples (see [36]). As a result, cyclic degradation must not be taken into account for high strength micro molded 3Y-TZP samples fatigued up to 1×10^5 cycles which fail due to small natural flaws similar in initial size to the dimension of the grain size. However, purely subcritical crack growth is relevant in this case, and a low toughness must be expected for such samples, because they fail before shielding can be developed.

In the literature, fracture analyses cannot be found for three-point-bending fatigue of ceramics at the size scale investigated in this work. This is the same for detailed micro molded aluminum bronze fatigue fracture analyses in the high cycle regime. Only few reports on macroscopic samples exist which can be compared to the fatigue fracture morphology of both micro molded materials. However, some features which are developed at the macroscopic level could not be found on the microscale (e.g. mirror on the 3Y-TZP fracture surface). Furthermore, small features, which are likely to be relevant at the microscale (e.g. micro cracks, micro cracks with extrusions or micro cracks covered with bulges of aluminum bronze), are not investigated or reported for macro samples, where the size of the microstructure and the crack lengths is at least one order of magnitude larger. Therefore, a catalog of micrographs from fatigued micro molded samples was established in this work. For future studies on the fatigue fracture morphology of micro molded materials, this catalog will help to categorize results and to evaluate discovered features. They can then be directly compared to the micro graphs of fatigued micro molded samples presented in the present work.

The same is valid for the collection of micro graphs showing the morphology of the aluminum bronze microstructure. Techniques such as EDX, EELS, and STEM-EDX were used to assign the precipitates which are visible in SEM images. Thus, it was found that lamellar κ_{III}-precipitates as well as dendritic κ_{II}-precipitates can be determined from their appearance in the SEM images. However, round-shaped precipitates can either be (degenerate lamellar) κ_{III}-precipitates or κ_{II}-precipitates, which cannot be distinguished in the SEM image. Furthermore, it was shown that the martensitic β phase can be revealed by ion milling in the dual beam microscope. The phases described by Hasan et al. [51] could be found in the samples, whereas their arrangement was denser and less ordered in the micro molded alloy. Moreover, the precipitates and grains tend to be about one order of magnitude smaller in micro molded samples than in macroscopic samples. Evidence for copper-rich precipitates inside lamellar κ_{III}-precipitates was found in the present work. Hasan et al. [51] observed copper-rich precipitates only inside κ_{I}-precipitates.

One major conclusion related to micro mechanical testing can be drawn: the size scale of all influencing factors has a remarkable impact on the determination of material properties. These factors are mainly the sample size, the microstructure and the testing equipment. At first sight, this statement might seem self-evident, but a detailed analysis shows that multiple consequences must be taken into account. They all must be kept in mind in order to design an adequate experiment and to interpret the results in the right way. Thus, technical components can be evaluated or the damage mechanism of a novel material can be elucidated.

Material parameters at the large scale are usually well known, as a continuum behavior can be found coming from an average response of statistically oriented and statistically arranged microstructural features. The continuum material behavior depends on the manufacturing process. For instance, rolled sheets are known to show a highly textured microstructure, with grains elongated and preferentially oriented along the rolling direction. In this case, the material parameters, such as Young's modulus, yield strength or tensile strength are anisotropic on the macroscopic scale.

6.2. Sample size

When the sample size of an isotropic material is scaled down, with the microstructure not being modified, the crystallographic anisotropy becomes visible in the overall sample behavior at one point. This is the case when too few grains or microstructural features are contained in the sample volume to represent an average distribution of all the microstructural variations of the material coming out of the manufacturing process. For instance, a sample at standard size (large scale), usually containing millions or billions of grains, will always contain grains oriented very close to the maximum Schmid factor of 0.5. This might not be the case any more for small samples, which contain only tens or hundreds of grains. As a result, such samples can show higher tensile strengths compared to large scale samples, when dislocation gliding is the failure inducing mechanism. The transition from the range of the macroscopic strength level to the range where an elevated strength can be observed is attributed to a size effect. Since the highest Schmid factor occurring in the grains of such small samples is expected to vary from sample to sample, the scatter of strength can be expected to be higher than for large scale samples, and to increase for decreasing sample sizes. This is a typical scaling effect, which was already treated in the first chapter of this document. The work of Furuya was mentioned as an important example for the fatigue behavior [27], showing larger scatter and higher lifetimes for smaller high strength steel samples. Here, the sample volume must be large enough to contain a representative population of failure inducing inclusions in order to show the homogeneous macroscopic behavior.

For industrial applications there might be an interest in extracting small samples from bulk material in order to manufacture small components (e.g. tiny screws, medical components such as stents, etc.). Furthermore, it might be easier or cheaper to test material properties at the small scale, and to extrapolate the results to the macroscopic scale. For instance, testing irradiated material is complex and expensive. The properties are important to design nuclear plants or fusion reactors, containing large components; however, small sample sizes are preferred to facilitate testing and manipulation of the material in radiation protected zones and to keep the amount of nuclear waste as small as possible. Also, for tests without radiation conducted on a large number of samples manufactured out of an expensive material (e.g. having a high content of copper), small sample sizes might lower the cost remarkably. Last but not least, local material properties of a large component can be of major interest. As an example, one can consider a small zone of stress concentration in the vicinity of a radius contained in a molded macroscopic part. This zone might be a critical site of failure initiation for the part, and therefore, of special interest. Since the microstructure in a molded part can vary locally, it might be necessary to extract small samples from the critical zone in order to investigate the phenomenon. A further example is the behavior of graded material. For instance, after hardening, the material properties change gradually from the surface towards the interior of a part. Also, for materials reinforced with particles or fibers, there might be a graded distribution of reinforcing elements related to the manufacturing process. As a result, the material behavior changes gradually from location to location. When the local properties shall be determined, there is often a need of cutting out small samples.

In all these cases, the relation between the sample size and the distribution of the critical microstructural features contained in the sample is crucial. It determines whether the sample is large enough for an extrapolation of the measured properties to the macroscopic scale. If this is not true, the scatter of the results is expected to be higher than for the macroscopic scale. One might try to find the macroscopic behaviour by averaging over a higher number of small scale tests. However, this can be a misleading attempt. The impact of large microstructural features might be missed when they do not fit entirely into the small sample size, and the interaction between microstructural features contained inside the sample might depend on the number and on the character of features, which are present in the tested volume. A good example is the work of Furuya [27]. Without knowing the results of large samples, it would be difficult to extrapolate from the small sample results to the larger scale. Thus, an extrapolation from the small scale to the large scale should not be done without validating experiments at both scales. For an extrapolation from the large scale to the small scale, the same reasoning can be applied.

When the local material parameters are of interest, the testing volume cannot be increased, but the points of discussion can help to interpret the generated data. A high overall scatter or unexpectedly high or low values for a material parameter at some site must not necessarily only be due to the local value of a graded property. It can also be (at least partially) a result of the reduced portion of microstructural variation, which is present inside the small sample volume. In most cases, it might be difficult or even impossible to separate such effects. To face this problem, the microstructure can be analyzed and evaluated with respect to the sample size using theoretical or numerical considerations. Furthermore, the amount of collected experimental information can be maximized analyzing different sample orientations and thicknesses, or different kinds of loading (tension, bending, ...). Since the graded material behavior and the scaling effect related to the sample size might be expected to affect such variations of the experimental conditions in different ways, a separation of the effects might become achievable. There is clear evidence that an investigation at the microscale can be much more elaborate compared to the macroscopic scale, when scaling effects are involved.

An important point, which was not discussed up to now, is the sample preparation. Manufacturing processes such as cutting operations usually affect a significant percentage of the sample volume at the micrometer scale (e.g. thermally affected zones from laser cutting or electro discharge machining, plastified zones from milling or grinding). For the preparation of micro samples, these influences should be minimized using soft process parameters, e.g. by means of cleaning cuts. Also etching techniques such as electro polishing might help, as long as an adequate etchant exists that attacks the sample surface uniformly. However, an eventually elevated influence of the sample preparation process on the mechanical properties must be considered for micro samples.

It is possible to avoid certain scaling effects by changing the manufacturing process. This was shown with the experiments conducted in the present work on aluminum bronze. The microstructure is scaled down in size with respect to macroscopic samples, leading to a bulk like fatigue behavior.

Thus, the sample volume of the micro molded sample is still large enough to contain a representative population of critical microstructural features leading to fatigue failure. As a result, the average fatigue behavior is comparable to the one of macroscopic samples, where the microstructure as well as the sample volume are larger. In contrast, if a micro sample were to be cut from a macroscopic aluminum bronze part having the coarse microstructure and casting defects on the order of micro samples, scaling effects would be expected to occur, leading to an elevated scatter of lifetimes. One can draw the conclusion that a bottom up manufacturing process such as molding or vapor deposition, which is optimized for the microscale, is preferable to produce micro parts. A more homogeneous behavior can be expected than for top down processes, such as milling or laser cutting from bulk material, which are more susceptible to scaling effects. This argument might be less valuable for bulk material having a very fine microstructure (compared to the micro sample size), but in this case still a relatively large zone affected by the cutting process might be an important risk.

6.3. Microstructure

When the mircrostructure is refined, e.g. in order to develop a bottom up process to manufacture micro parts, microstructural scaling effects and size effects must be taken into account. An example for a scaling effect related purely to the microstructure and not to the sample size is the Hall-Petch effect. When the grain size is reduced, the yield strength increases. Another microstructural scaling effect was found in the present work for the zirconia samples. Here, the defect sizes in the samples are scaled down. As a result, the crack extension arising from a natural flaw during fatigue ends up to be so small that no degradable shielding can develop and degradation is purely related to subcritical crack growth. Since the degradation of shielding can be interpreted as a mechanism, which is suppressed at a certain size scale, one could also assign the phenomenon to the size effects. On the other hand, the persisting degradation mechanism, namely subcritical crack growth, also exists on the larger scale in addition to cyclic degradation of shielding. Thus, no switch to a new failure mechanism occurs at the small scale, giving reason to classify the phenomenon as a microstructural scaling effect. Clear size effects, where a new deformation mechanism appears due to a refined microstructure, arise in thin films, nanocrystalline materials and nanotwinned materials. It is very important to consider such effects, when the mechanical behavior of micro samples is investigated. This is crucial for the development of micro sample manufacturing processes, leading to a refined microstructure. When size and scaling effects are not considered, experimental results can be misinterpreted. Furthermore, important investigations might be omitted, because they are only relevant when size effects can occur.

6.4. Experimental equipment

The discussion above shows a clear need for mechanical testing on the microscale, in order to take size and scaling effects into account. Testing equipment for the macroscopic size scale with standard sample cross sections in the range of about $7\,\mathrm{mm}^2$ to $540\,\mathrm{mm}^2$ is well developed and commercially available (the cross sections are calculated from the range of dimensions mentioned in the German standard DIN 50125 for the geometry of metallic tensile samples). However, commercially available test rigs for the microscale are missing and tests at the microscale are barely standardized. The main challenge is to reach sufficiently accurate resolutions for displacement, strain and load measurements, while providing a large enough measurement range. For instance, when the dimensions of a tensile sample are scaled down linearly, the load, which is needed to apply a certain stress, scales down quadratically. Therefore, the load resolution must rise disproportionately. This can typically be realized by reducing the stiffness of the load cell. However, such a design brings some disadvantages. A more compliant load cell typically reduces the natural frequency of a dynamic test rig and therefore limits the achievable testing frequency. Furthermore, the amount of elastic energy stored in a compliant load cell is elevated. This energy can be released when the tested material deforms, degrading the quality of the load or displacement control from the experiment.

When the sample is scaled down, the applied loads come closer and closer to the level of frictional, adhesive, or inertial loads which arise in the testing device. This can be challenging, e.g., for the three-point-bending device shown in the present work, where a special low friction support had to be designed. Furthermore, a free hanging mass without guidance was chosen for the resonant system to avoid the influence of friction. The x-direction motion of the tensile setup was not supported transversally beyond the piezo's internal guidance. This was in order to maintain the high stiffness and the highly dynamic behavior of the device, even though it demands some compromise in terms of lateral accuracy. On the other hand, new opportunities are provided by mechanical tests conducted at a smaller scale. For instance, small samples can sometimes just be glued into grips, or masses determining the resonant frequency of a sample can be attached using adhesives. A mass of 30 g, which is easy to manipulate, was sufficient to adjust the natural frequency of the micro molded samples in the present work. Self heating did not compromise the fatigue tests in the kHz-regime, due to the high surface to volume ratio of the samples. Furthermore, it was possible to detect cracks of only several tens of micrometers length by means of the load or displacement signal, because they reduce the stiffness of the micro samples remarkably. The testing devices are compact and relatively economically priced, once they are developed.

The present work shows promising ways of testing micro samples in tension, compression and bending, even cyclically or quasi-statically. In the future, strain controlled experiments as well as an extension of the methods to high and low temperatures shall be realized. Since more and more material data at the microscale is needed for industrial applications, it would be important to develop testing standards for this size scale, based on testing methods as they were developed in the present work. This would open the way to develop reliable standardized design and testing methods for small components, such as miniaturized sensors, thin foils or parts of micro mechanical systems (MEMS-devices). These devices are contained in multiple products that we are using in our daily lives, e.g. in cars, mobile phones, computers, medical instruments, electric motors, wind mills, et cetera.

A. Appendix

sample	legend	cycles	σ_a [MPa]	run out	f [Hz]	R	width [µm]	thick. [µm]	cross sec. [µm²]	FIB
2229-5	25 Hz,R=-0.1	8.40×10^3	360	-	25	0.1	267	121	0.032307	-
2229-7		5.55×10^4	291	-	25	0.1	271	121	0.032791	-
2229-10		2.92×10^5	200	-	25	0.1	257	130	0.03341	fib
2229-11		2.13×10^5	219	-	25	0.1	265	128	0.03392	-
2229-13		1.64×10^4	333	-	25	0.1	264	126	0.033264	-
2233-6	100Hz,R=-0.1	2.90×10^5	180	-	100	0.1	263	120	0.03156	fib
2317-4	900Hz,R=-0.9	4.57×10^5	250	-	868	-0.9	262	120	0.03144	-
2233-3		4.91×10^5	222	-	902	-0.9	249	115	0.028635	fib
2317-8		4.30×10^8	170	run out	897	-0.88	267	120	0.03204	-
2317-6		1.68×10^6	190	-	927	-0.89	266	118	0.031388	fib
2319-2		1.58×10^9	200	run out	932	-0.89	251	110	0.02761	fib
2318-8	1100Hz,R=-0.9	1.09×10^9	180	run out	1079	-0.89	262	113	0.029606	fib
2318-15		5.56×10^8	190	run out	1078	-0.9	266	118	0.031388	-
2314-15		5.62×10^5	220	-	1059	-0.92	264	126	0.033264	-
2318-10	1000Hz,R=-0.9	6.90×10^5	200	-	995	-0.9	255	115	0.029325	fib
2318-2		1.42×10^7	200	-	1035	-0.9	264	118	0.031152	fib
2318-4		3.99×10^5	200	-	992	-0.9	259	118	0.030562	-
2318-9		7.93×10^8	200	run out	996	-0.9	263	120	0.03156	fib
2319-15		4.80×10^9	170	run out	1008	-0.88	261	115	0.030015	-
2319-14		5.20×10^5	220	-	952	-0.9	254	113	0.028702	-
2318-16		4.35×10^8	215	-	1033	-0.9	261	113	0.029493	-
2317-14	2200Hz,R=-1	1.27×10^9	220	run out	2154	-0.98	265	120	0.0318	-
2318-6	2400Hz,R=-1	3.98×10^5	220	-	2362	-0.98	116	258	0.029928	fib

Table A.1. Fatigue data of **as-cast** micro molded aluminum bronze as shown in figure 4.5. The bold part of the sample name indicates the batch number, the consecutive part is the sample number designating the sample position on the cast part. The column "legend" refers to the graph in figure 4.5.

sample	legend	cycles	σ_a [MPa]	run out	f [Hz]	R	width [µm]	thick. [µm]	cross sec. [µm^2]	FIB
2319-9	2.5Hz,R=0.1	4.08×10^4	320	-	2.5	0.1	258	60	0.01548	-
2319-13	1000Hz,R=-0.9	2.10×10^6	210	-	1007	-0.9	259	76	0.019684	fib
2319-9TW		3.50×10^6	190	-	1005	-0.89	266	70	0.01862	-
2319-7	1100Hz,R=-0.9	3.60×10^7	228	-	1142	-0.93	265	96	0.02544	fib
2314-8		1.05×10^9	190	run out	1056	-0.91	262	92	0.024104	-
2314-11	1200Hz,R=-0.9	2.32×10^9	180	run out	1202	-0.92	264	105	0.02772	-
2315-5		1.12×10^9	200	run out	1212	-0.93	261	109	0.028449	-
2314-11A	100Hz,R=0.1	7.16×10^7	160	run out	100	0.1	262	84	0.022008	-

Table A.2. Fatigue data of **polished** micro molded aluminum bronze as shown in figure 4.5. The bold part of the sample name indicates the batch number, the consecutive part is the sample number designating the sample position on the cast part. The column “legend” refers to the graph in figure 4.5.

batch F4-3_A6_S6

sample	σ_a [MPa]	cycles	run out/ spon. frac.	width [µm]	thickness [µm]	frequency Hz
5	233	1.676×10^3	-	192	205	25
6	390	1	s	195	198	25
7	381	1	s	200	197	25
8	380	$3.809\,99 \times 10^5$	run out	196	192	25
	391	1.457×10^5	run out			25
	391	2.379×10^5	run out			75
	391	1.532×10^5	run out			100
	401	1.722×10^5	run out			100
	411	5.0×10^2	-			100
9	374	1.1×10^3	-	192	197	100

Table A.3. Samples of batch F4-3_A6_S6, which were fatigued at different frequencies and amplitudes with a load ratio of $R = 0.2$ for pretests. Spontaneous failures before attaining the maximum load are designated with "s" in the column "run out". Sample number 8 is listed with multiple parameters. In this case, subsequent tests (at higher loads or frequencies) were conducted, after a high number of cycles was attained without fracture. The order in the table corresponds to the order in which these tests were conducted.

batch F4-3_A6_S6

sample	σ_a [MPa]	cycles	run out/ spon. frac.	width [μm]	thickness [μm]
8	>320	$>7.0 \times 10^8$	run out	196	192
10	327	1.3699×10^4	-	193	196
11	321	3.6909×10^4	-	200	192
12	306	1.2999×10^4	-	200	194
13	324	1.000×10^3	-	202	195
14	324	4.0149×10^4	-	194	199
15	318	1.700×10^3	-	192	200
16	322	2.4872099×10^7	run out	195	199
17	320	1	s	195	197
18	317	1	s	199	193
19	315	5.150×10^3	-	195	196
20	314	1	s	191	199
21	320	6.450×10^3	-	192	201
22	320	2.250×10^3	-	198	192
23	320	5.9898×10^4	-	195	200
24	320	1	s	196	202
25	320	1	s	193	198
26	320	1.650×10^3	-	194	198
27	319	6.400×10^3	-	195	204
28	320	1.07449×10^5	-	201	191

Table A.4. Samples of batch F4-3_A6_S6, which were fatigued at 100 Hz with a load ratio of $R = 0.2$. Spontaneous failures before attaining the maximum load are designated with "s" in the column "run out". The data of these samples (including number 8 from pretests) were used to calculate the crack propagation exponents.

batch **F2**-2_A10_S10

sample	σ_a [MPa]	cycles	run out/ spon. frac.	width [µm]	thickness [µm]
1	381	$8.107\,399 \times 10^6$	run out	200	207
	400	1.088×10^6	run out		
	420	4.63×10^5	run out		
	440	4.75×10^5	run out		
	460	1.0×10^5	run out		
	480	1.4×10^5	run out		
	500	1.0×10^6	run out		
	520	1.0×10^4	-		
2	500	$2.487\,99 \times 10^5$	-	213	203
3	521	8.3465×10^5	run out	208	200
	542	1.727×10^5	run out		
	562	1.35×10^5	run out		
	583	1.38×10^5	-		
4	560	5.7×10^2	-	209	199
5	560	6.15×10^2	-	201	207
6	541	1	-	201	212
7	540	1	-	211	201
8	520	$2.668\,809\,9 \times 10^7$	run out	208	204
	540	1.1×10^5	run out		
	560	9.0×10^4	run out		
	580	1.0×10^5	run out		
	600	1.1×10^5	run out		
	620	7.0×10^4	run out		
	640	1.0×10^4	-		
9	541	1	s	200	210
10	501	$1.613\,49 \times 10^5$	-	203	207
11	502	1	s	212	203
12	500	8.849×10^3	-	210	201
13	500	8.5058×10^6	run out	200	209
	540	7.96×10^5	run out		
	580	6.0×10^3	-		
14	500	$7.718\,549 \times 10^6$	run out	202	209
	540	7.61×10^5	run out		
	580	$9.070\,09 \times 10^6$	run out		
	620	9.5×10^5	run out		
	660	2.2×10^5	-		
15	500	2.75×10^3	-	209	201

Table A.5. Samples of batch F2-2_A10_S10, which were fatigued at 100 Hz with a load ratio of $R = 0.2$. Spontaneous failures before attaining the maximum load are designated with "s" in the column "run out". The number of cycles to failure is noted in the column "cycles". Subsequent tests (at higher loads) were conducted with samples for which a run out occurred. In these cases, the specific numbers of cycles, after which the load was increased, are noted in the column "cycles" of the table.

batch F2-2_A10_S10

sample	σ_a **[MPa]**	**cycles**	**run out/ spon. frac.**	**width [µm]**	**thickness [µm]**
16	500	$6.688\,098 \times 10^6$	-	210	204
17	501	1.1×10^3	-	209	203
18	500	1.4×10^3	-	210	201
19	500	$7.998\,449 \times 10^6$	run out	209	200
	540	$8.594\,535 \times 10^6$	run out		
	580	$1.873\,731 \times 10^6$	run out		
	620	6.621×10^6	run out		
	660	4.7234×10^4	-		
20	500	8.85×10^3	-	207	201
21	500	$1.225\,537 \times 10^6$	run out	197	208
	540	1.037×10^6	run out		
	580	1.0584×10^4	-		
22	500	4.65×10^3	-	211	203
23	500	6.4×10^3	-	201	210
24	501	1	s	202	208
25	500	4.05×10^3	-	199	207
26	501	1	s	210	201
27	501	1	s	200	210
28	500	$1.517\,057 \times 10^6$	run out	201	212
	540	$2.292\,557 \times 10^6$	-		
29	501	1	s	210	200

Table A.6. Samples of batch F2-2_A10_S10, which were fatigued at 100 Hz with a load ratio of $R = 0.2$. Spontaneous failures before attaining the maximum load are designated with "s" in the column "run out". The number of cycles to failure is noted in the column "cycles". Subsequent tests (at higher loads) were conducted with samples, for which a run out occurred. In these cases, the specific numbers of cycles, after which the load was increased, are noted in the column "cycles" of the table.

batch F11-4_A2_S2

sample	σ_a [MPa]	**cycles**	**run out/ spon. frac.**	**width** [µm]	**thickness** [µm]
2	801	1.1550×10^3	-	192	197
3	800	7.0950×10^4	-	196	201
4	784	2.2530×10^4	-	196	200
5	800	1.4510×10^3	-	196	201
6	0	1	s	196	201
7	800	1.2010×10^3	-	188	196
8	801	1.7350×10^4	-	194	199
9	0	1	s	198	201
10	800	4.2510×10^3	-	199	195
11	800	2.6010×10^3	-	199	210
12	801	1	s	207	197
13	800	3.5010×10^3	-	206	199
14	801	5.5000×10^2	-	208	200
15	800	1	s	208	200
16	800	4.6600×10^2	-	210	201

Table A.7. Samples of batch F11-4_A2_S2, which were fatigued at 100 Hz with a load ratio of $R = 0.2$. Spontaneous failures before attaining the maximum load are designated with "s" in the column "run out". The data of these samples (including number 8 from pretests) were used to calculate the crack propagation exponents.

batch F2-2_A10_S10		**batch F4**-3_A6_S6		**batch F11**-4_A2_S2	
amplitude $\sigma_a = 500\,\text{MPa}$		amplitude $\sigma_a = 320\,\text{MPa}$		amplitude $\sigma_a = 800\,\text{MPa}$	
cycles	**sample**	**cycles**	**sample**	**cycles**	**sample**
1.1×10^3	17	1.0×10^3	13	4.66×10^2	16
1.4×10^3	18	1.65×10^3	26	5.50×10^2	14
2.75×10^3	15	1.7×10^3	15	1.155×10^3	2
4.05×10^3	25	2.25×10^3	22	1.201×10^3	7
4.65×10^3	22	5.15×10^3	19	1.451×10^3	5
6.4×10^3	23	6.4×10^3	27	2.601×10^3	11
8.849×10^3	12	6.45×10^3	21	3.501×10^3	13
8.85×10^3	20	1.2999×10^4	12	4.251×10^3	10
$1.613\,49 \times 10^5$	10	1.3699×10^4	10	1.7350×10^4	8
$2.487\,99 \times 10^5$	2	3.6909×10^4	11	2.2530×10^4	4
run out	1	4.0149×10^4	14	7.095×10^4	3
run out	3	5.9898×10^4	23	s	6
run out	8	$1.074\,49 \times 10^5$	28	s	9
run out	13	run out	8	s	12
run out	14	run out	16	s	15
run out	16	s	17		
run out	19	s	18		
run out	21	s	20		
s	11	s	24		
s	24	s	25		
s	26				
s	27				
s	29				

Table A.8. Fatigue data, which were used to generate the crack growth curves. The lifetimes as well as the number of run outs and of spontaneous fractures (designated with "s") are needed for the statistical analysis. Further information contained in this analysis: R-ratio $R = 0.2$, fracture toughness $K_{IC} = 4\,\text{MPam}^{1/2}$, geometry factor $Y_\pi = 1.3$, strength data of the manufacturer.

Bibliography

[1] L. C. A. Strange, "On aluminium bronze as a material for the construction of astronomical and other philosophical instruments," *Journal of the Franklin Institute*, vol. 75, no. 6, pp. 414 – 419, 1863.

[2] H. Meigh, *Cast and wrought aluminium bronzes properties, processes and structure.* IOM Communications, London, 2000.

[3] Deutsches Kupferinstitut, "Werkstoff-Datenblatt CuAl10Fe5Ni5-C - CC333G (2.0975.01, .02, .03, .04)," 2005. http://www.kupferinstitut.de/front_frame/pdf/CuAl10Fe5Ni5-C.pdf (3/4/2013).

[4] R. C. Garvie, R. H. Hannink, and R. T. Pascoe, "Ceramic steel?," *Nature*, vol. 258, no. 5537, pp. 703 – 704, 1975.

[5] D. Munz and T. Fett, *Ceramics, mechanical properties, failure behaviour, materials selection.* Springer-Verlag Berlin Heidelberg, 1999.

[6] J. R. Kelly and I. Denry, "Stabilized zirconia as a structural ceramic: an overview.," *Dental materials official publication of the Academy of Dental Materials*, vol. 24, no. 3, pp. 289 – 298, 2008.

[7] J. Chevalier, L. Gremillard, A. V. Virkar, and D. R. Clarke, "The tetragonal-monoclinic transformation in zirconia: Lessons learned and

future trends," *Journal of the American Ceramic Society*, vol. 92, no. 9, pp. 1901 – 1920, 2009.

[8] S. Suresh, *Fatigue of materials.* Cambridge University Press, 1991.

[9] H. Mughrabi, "Fatigue, an everlasting materials problem - still en vogue," *Procedia Engineering*, vol. 2, no. 1, pp. 3 – 26, 2010. Fatigue 2010.

[10] H. Mughrabi, "Cyclic slip irreversibilities and the evolution of fatigue damage," *Metallurgical and Materials Transactions B*, vol. 40, pp. 431 – 453, 2009.

[11] C. Sonsino, "Course of SN-curves especially in the high-cycle fatigue regime with regard to component design and safety," *International Journal of Fatigue*, vol. 29, no. 12, pp. 2246 – 2258, 2007.

[12] Y. Murakami, N. N. Yokoyama, and J. Nagata, "Mechanism of fatigue failure in ultralong life regime," *Fatigue & Fracture of Engineering Materials & Structures*, vol. 25, no. 8-9, pp. 735 – 746, 2002.

[13] T. Sakai, "Review and prospects for current studies on very high cycle fatigue of metallic materials for machine structural use," *Journal of Solid Mechanics and Materials Engineering*, vol. 3, no. 3, pp. 425 – 439, 2009.

[14] K. Shiozawa, Y. Morii, S. Nishino, and L. Lu, "Subsurface crack initiation and propagation mechanism in high-strength steel in a very high cycle fatigue regime," *International Journal of Fatigue*, vol. 28, no. 11, pp. 1521 – 1532, 2006.

[15] H. Mughrabi, "Specific features and mechanisms of fatigue in the ultrahigh-cycle regime," *International Journal of Fatigue*, vol. 28, no. 11, pp. 1501 – 1508, 2006.

[16] C. Bathias, "Piezoelectric fatigue testing machines and devices," *International Journal of Fatigue*, vol. 28, no. 11, pp. 1438 – 1445, 2006.

[17] U. Krupp, *Fatigue crack propagation in metals and alloys.* Wiley-VCH Verlag GmbH & Co. KGaA, Weinheim, 2007.

[18] J. Schijve, *Fatigue of structures and materials.* Springer Science+Business Media, B.V., 2009.

[19] K.-H. Grote and J. Feldhusen, eds., *Dubbel : Taschenbuch für den Maschinenbau.* Berlin Heidelberg: Springer-Verlag, 22., neubearb. und erw. Aufl. ed., 2007.

[20] D. Taylor and J. F. Knott, "Fatigue crack propagation behaviour of short cracks; the effect of microstructure," *Fatigue & Fracture of Engineering Materials & Structures*, vol. 4, no. 2, pp. 147 – 155, 1981.

[21] D. Taylor and J. F. Knott, "Growth of fatigue cracks from casting defects in nickel-aluminum bronze," *Metals Technology*, vol. 9, p. 221, 1982.

[22] Mughrabi, "On the life-controlling microstructural fatigue mechanisms in ductile metals and alloys in the gigacycle regime," *Fatigue & Fracture of Engineering Materials & Structures*, vol. 22, no. 7, pp. 633 – 641, 1999.

[23] H. Mayer and C. Laird, "Influence of cyclic frequency on strain localization and cyclic deformation in fatigue," *Materials Science and Engineering: A*, vol. 187, no. 1, pp. 23 – 35, 1994.

[24] H. Mayer and C. Laird, "Frequency effects on cyclic plastic strain of polycrystalline copper under variable loading," *Materials Science and Engineering: A*, vol. 194, pp. 137 – 145, 1995.

[25] M. Papakyriacou, H. Mayer, C. Pypen, H. Plenk Jr., and S. Stanzl-Tschegg, "Influence of loading frequency on high cycle fatigue properties of b.c.c. and h.c.p. metals," *Materials Science and Engineering: A*, vol. 308, no. 1–2, pp. 143 – 152, 2001.

[26] H. Mayer, M. Papakyriacou, R. Pippan, and S. Stanzl-Tschegg, "Influence of loading frequency on the high cycle fatigue properties of AlZnMgCu1.5 aluminium alloy," *Materials Science and Engineering: A*, vol. 314, no. 1–2, pp. 48 – 54, 2001.

[27] Y. Furuya, "Specimen size effects on gigacycle fatigue properties of high-strength steel under ultrasonic fatigue testing," *Scripta Materialia*, vol. 58, no. 11, pp. 1014 – 1017, 2008.

[28] S. Stanzl-Tschegg, H. Mughrabi, and B. Schoenbauer, "Life time and cyclic slip of copper in the VHCF regime," *International Journal of Fatigue*, vol. 29, no. 9–11, pp. 2050 – 2059, 2007.

[29] S. E. Stanzl-Tschegg and B. Schönbauer, "Mechanisms of strain localization, crack initiation and fracture of polycrystalline copper in the VHCF regime," *International Journal of Fatigue*, vol. 32, no. 6, pp. 886 – 893, 2010.

[30] Y. Murakami, T. Nomoto, T. Ueda, and Y. Murakami, "On the mechanism of fatigue failure in the superlong life regime ($n>10^7$ cycles). Part I: influence of hydrogen trapped by inclusions," *Fatigue & Fracture of Engineering Materials & Structures*, vol. 23, no. 11, pp. 893 – 902, 2000.

[31] Y. Murakami, T. Nomoto, T. Ueda, and Y. Murakami, "On the mechanism of fatigue failure in the superlong life regime ($n>10^7$ cycles). Part II: influence of hydrogen trapped by inclusions," *Fatigue*

& *Fracture of Engineering Materials & Structures*, vol. 23, no. 11, pp. 903 – 910, 2000.

[32] A. McEvily, T. Nakamura, H. Oguma, K. Yamashita, H. Matsunaga, and M. Endo, "On the mechanism of very high cycle fatigue in Ti-6Al-4V," *Scripta Materialia*, vol. 59, no. 11, pp. 1207 – 1209, 2008.

[33] R. Schwaiger and O. Kraft, "Size effects in the fatigue behavior of thin Ag films," *Acta Materialia*, vol. 51, no. 1, pp. 195 – 206, 2003.

[34] D. Gianola, C. Eberl, X. Cheng, and K. Hemker, "Stress-driven surface topography evolution in nanocrystalline Al thin films," *Advanced Materials*, vol. 20, pp. 303 – 308, 2008.

[35] G. Richter, K. Hillerich, D. S. Gianola, R. Mönig, O. Kraft, and C. A. Volkert, "Ultrahigh strength single crystalline nanowhiskers grown by physical vapor deposition," *Nano Letters*, vol. 9, no. 8, pp. 3048 – 3052, 2009.

[36] J. Chevalier, C. Olagnon, and G. Fantozzi, "Subcritical crack propagation in 3Y-TZP ceramics: Static and cyclic fatigue," *Journal of the American Ceramic Society*, vol. 82, no. 11, pp. 3129 – 3138, 1999.

[37] L. Ewart and S. Suresh, "Dynamic fatigue crack growth in polycrystalline alumina under cyclic compression," *Journal of Materials Science Letters*, vol. 5, no. 8, pp. 774 – 778, 1986.

[38] R. H. Dauskardt, W. Yu, and R. O. Ritchie, "Fatigue crack propagation in transformation-toughened zirconia ceramic," *Journal of the American Ceramic Society*, vol. 70, no. 10, pp. C–248 – C–252, 1987.

[39] S.-Y. Liu and I.-W. Chen, "Fatigue of yttria-stabilized zirconia: I, fatigue damage, fracture origins, and lifetime prediction," *Journal of the American Ceramic Society*, vol. 74, no. 6, pp. 1197 – 1205, 1991.

[40] S.-Y. Liu and I.-W. Chen, "Fatigue of yttria-stabilized zirconia: II, crack propagation, fatigue striations, and short-crack behavior," *Journal of the American Ceramic Society*, vol. 74, no. 6, pp. 1206 – 1216, 1991.

[41] M. J. Hoffman, W. Lentz, M. V. Swain, and Y.-W. Mai, "Cyclic fatigue lifetime predictions of partially stabilized zirconia with crack resistance curve characteristics," *Journal of the European Ceramic Society*, vol. 11, no. 5, pp. 445 – 453, 1993.

[42] S. Lathabai, J. Rödel, and B.-R. Lawn, "Cyclic fatigue from frictional degradation at bridging grains in alumina," *Journal of the American Ceramic Society*, vol. 74, no. 6, pp. 1340 – 1348, 1991.

[43] D. S. Jacobs and I.-W. Chen, "Cyclic fatigue in ceramics: A balance between crack shielding accumulation and degradation," *Journal of the American Ceramic Society*, vol. 78, no. 3, pp. 513 – 520, 1995.

[44] S.-Y. Liu and I.-W. Chen, "Fatigue deformation mechanisms of zirconia ceramics," *Journal of the American Ceramic Society*, vol. 75, no. 5, pp. 1191 – 204, 1992.

[45] T. Fett and D. Munz, "Influence of time-dependent phase transformations on bending tests," *Materials Science and Engineering A*, vol. 219, no. 1-2, pp. 89 – 94, 1996.

[46] T. R. Finlayson, A. K. Gross, J. R. Griffiths, and E. H. Kisi, "Creep of Mg-PSZ at room temperature," *Journal of the American Ceramic Society*, vol. 77, no. 3, pp. 617 – 624, 1994.

[47] R. Schmitt, T. Fett, and D. Munz, "Cyclic fatigue of zirconia," *Fatigue & Fracture of Engineering Materials & Structures*, vol. 19, no. 12, pp. 1411 – 1420, 1996.

[48] T. Fett and D. Munz, "Subcritical crack growth of macrocracks in alumina with R-curve behavior," *Journal of the American Ceramic Society*, vol. 75, no. 4, pp. 958 – 963, 1992.

[49] T. Fett, G. Martin, D. Munz, and G. Thun, "Determination of $da/dN - \Delta K_1$ curves for small cracks in alumina in alternating bending tests," *Journal of Materials Science*, vol. 26, pp. 3320 – 3328, 1991.

[50] M. Härtelt, *Probabilistische Lebensdauervorhersage für keramische Bauteile unter komplexer zyklischer Beanspruchung.* PhD thesis, Karlsruhe Institute of Technology (KIT), 2011.

[51] F. Hasan, A. Jahanafrooz, G. Lorimer, and N. Ridley, "The morphology, crystallography, and chemistry of phases in as-cast nickel-aluminum bronze," *Metallurgical and Materials Transactions A*, vol. 13, pp. 1337 – 1345, 1982.

[52] P. Weill-Couly and D. Arnaud, "Influence de la composition et de la structure des cupro-aluminiums sur leur comportement en service," *Fonderie*, vol. 322, pp. 123 – 135, 1973.

[53] G. Baumeister, R. Ruprecht, and J. Haußelt, "Microcasting of parts made of metal alloys," *Microsystem Technologies*, vol. 10, no. 3, pp. 261 – 264, 2004.

[54] G. Baumeister, B. Okolo, J. Rögner, R. Ruprecht, E. Kerscher, V. Schulze, J. Haußelt, and D. Löhe, "New results on microcasting of Al bronze," *Microsystem Technologies*, vol. 14, pp. 1813 – 1821, 2008.

[55] J. Rögner, *Mechanische Eigenschaften urgeformter Mikroproben aus CuAl10Ni5Fe4, ZrO2 und Si3N4.* PhD thesis, Karlsruhe Institute of Technology (KIT), 2010.

[56] P. F. Becher and M. V. Swain, "Grain-size-dependent transformation behavior in polycrystalline tetragonal zirconia," *Journal of the American Ceramic Society*, vol. 75, no. 3, pp. 493 – 502, 1992.

[57] T. R. Lai, C. L. Hogg, and M. V. Swain, "Evaluation of fracture toughness and R-curve behaviour of Y-TZP ceramics," *ISIJ International*, vol. 29, no. 3, pp. 240 – 245, 1989.

[58] R. Matt, *Statisches und zyklisches Ermüdungsverhalten umwandlungsverstärkter ZrO_2-Werkstoffe*. PhD thesis, Universität Karlsruhe, 1996.

[59] L. Pan and S. Horibe, "An in-situ investigation on the critical phase transformation stress of tetragonal zirconia polycrystalline ceramics," *Journal of Materials Science*, vol. 31, pp. 6523 – 6527, 1996.

[60] M. Matsuzawa, E. Fujimagari, and S. Horibe, "Cyclic deformation and crack growth in zirconia ceramics," *Materials Science and Engineering: A*, vol. 314, no. 1–2, pp. 105 – 109, 2001.

[61] V. Piotter, W. Bauer, T. Benzler, and A. Emde, "Injection molding of components for microsystems," *Microsystem Technologies*, vol. 7, no. 3, pp. 99 – 102, 2001.

[62] F. Çetinel, W. Bauer, R. Knitter, and J. Haußelt, "Factors affecting strength and shape retention of zirconia micro bending bars during thermal debinding," *Ceramics International*, vol. 37, no. 7, pp. 2809 – 2820, 2011.

[63] W. N. Sharpe, Jr. and J. Bagdahn, "Fatigue testing of polysilicon–a review," *Mechanics of Materials*, vol. 36, no. 1-2, pp. 3 – 11, 2004. Fatigue of Advanced Materials.

[64] D. Alsem, O. Pierron, E. Stach, C. Muhlstein, and R. Ritchie, "Mechanisms for fatigue of micron-scale silicon structural films," *Advanced Engineering Materials*, vol. 9, no. 1-2, pp. 15 – 30, 2007.

[65] H. Kahn, R. Ballarini, and A. Heuer, "Dynamic fatigue of silicon," *Current Opinion in Solid State and Materials Science*, vol. 8, no. 1, pp. 71 – 76, 2004.

[66] B. L. Boyce, J. R. Michael, and P. G. Kotula, "Fatigue of metallic microdevices and the role of fatigue-induced surface oxides," *Acta Materialia*, vol. 52, no. 6, pp. 1609 – 1619, 2004.

[67] O. Kraft, R. Schwaiger, and P. Wellner, "Fatigue in thin films: lifetime and damage formation," *Materials Science and Engineering: A*, vol. 319–321, pp. 919 – 923, 2001.

[68] C. Eberl, R. Spolenak, E. Arzt, F. Kubat, A. Leidl, W. Ruile, and O. Kraft, "Ultra high-cycle fatigue in pure Al thin films and line structures," *Materials Science and Engineering: A*, vol. 421, no. 1–2, pp. 68 – 76, 2006.

[69] B. Yang, C. Motz, W. Grosinger, W. Kammrath, and G. Dehm, "Tensile behaviour of micro-sized copper wires studied using a novel fibre tensile module," *International Journal of Materials Research*, vol. 07, pp. 716 – 724, 2008.

[70] B. Yang, C. Motz, W. Grosinger, and G. Dehm, "Stress-controlled fatigue behaviour of micro-sized polycrystalline copper wires," *Materials Science and Engineering: A*, vol. 515, no. 1–2, pp. 71 – 78, 2009.

[71] D. Kiener, C. Motz, W. Grosinger, D. Weygand, and R. Pippan, "Cyclic response of copper single crystal micro-beams," *Scripta Materialia*, vol. 63, no. 5, pp. 500 – 503, 2010.

[72] M. Funk, *Microstructural stability of nanostructured fcc metals during cyclic deformation and fatigue.* PhD thesis, Karlsruhe Institute of Technology (KIT), 2012.

[73] G. Khatibi, J. Horky, B. Weiss, and M. Zehetbauer, "High cycle fatigue behaviour of copper deformed by high pressure torsion," *International Journal of Fatigue*, vol. 32, no. 2, pp. 269 – 278, 2010.

[74] G. Khatibi, A. Betzwar-Kotas, V. Gröger, and B. Weiss, "A study of the mechanical and fatigue properties of metallic microwires," *Fatigue & Fracture of Engineering Materials & Structures*, vol. 28, no. 8, pp. 723 – 733, 2005.

[75] W. N. Sharpe, Jr., "Fourth international conference on very high cycle fatigue," 2007.

[76] W. N. Sharpe, Jr., "The interferometric strain gage," *Experimental Mechanics*, vol. 8, pp. 164 – 170, 1968.

[77] W. N. Sharpe, Jr., "Final report high-frequency high-temperature strain/displacement gage," Tech. Rep. DARPA Contract No: FA8650-04-5212, Johns Hopkins University, Baltimore, 2007.

[78] C. Eberl *et al.*, "Digital image correlation and tracking (matlab functions, online available)," 2010. http://www.mathworks.com/matlabcentral/fileexchange/12413-digital-image-correlation-and-tracking (10/12/2013).

[79] D. Gross, W. Hauger, W. Schnell, and P. Wriggers, *Technische Mechanik : Band 4: Hydromechanik, Elemente der Höheren Mechanik, Numerische Methoden.* Springer-Verlag Berlin Heidelberg, 6., vollständig neu bearbeitete Auflage ed., 2007.

[80] K. Hemker and W. N. Sharpe, Jr., "Microscale characterization of mechanical properties," *Annual Review of Materials Research*, vol. 37, no. 1, pp. 93 – 126, 2007.

[81] N. J. Briot, T. Kennerknecht, C. Eberl, and T. J. Balk, "Mechanical properties of bulk single crystalline nanoporous gold investigated by millimetre-scale tension and compression testing," *Philosophical Magazine*, accepted November 2013.

[82] S. Zhao, *Multiscale modeling of thermomechanical properties of ceramic pebbles.* PhD thesis, Karlsruhe Institute of Technology (KIT), 2010.

[83] S. Zhao, Y. Gan, M. Kamlah, T. Kennerknecht, and R. Rolli, "Influence of plate material on the contact strength of Li_4SiO_4 pebbles in crush tests and evaluation of the contact strength in pebble–pebble contact," *Engineering Fracture Mechanics*, vol. 100, pp. 28 – 37, 2013.

[84] T. Bückmann, N. Stenger, M. Kadic, J. Kaschke, A. Frölich, T. Kennerknecht, C. Eberl, M. Thiel, and M. Wegener, "Tailored 3D mechanical metamaterials made by dip-in direct-laser-writing optical lithography," *Advanced Materials*, vol. 24, no. 20, pp. 2710 – 2714, 2012.

[85] S. Slaby, "Mechanische Eigenschaften von elektrochemisch abgeschiedenen nanokristallinen Nickel-Wolfram-Schichten," Master's thesis, Karlsruhe Institue of Technology (KIT), 2011.

[86] Physik Instrumente (PI) GmbH & Co. KG, *Grundlagen der Nanostelltechnik*, cat120d inspirationen2009 09/02.10 ed., 2009. http://www.physikinstrumente.de/de/pdf_extra/2009_PI_Katalog_Grundlagen_der_Nanostelltechnik-Tutorial.pdf (11/21/2011).

[87] I. N. Bronstein, K. A. Semendjajev, G. Musiol, and H. Mühlig, *Taschenbuch der Mathematik.* Verlag Harri Deutsch, Frankfurt am Main Thun, 4., neubearb. und erw. Aufl. ed., 1999.

[88] D. Tabor, *The hardness of metals.* Clarendon press, Oxford, 1951.

[89] A. Atkins and D. Tabor, "Plastic indentation in metals with cones," *Journal of the Mechanics and Physics of Solids*, vol. 13, no. 3, pp. 149 – 164, 1965.

[90] A. C. Fischer-Cripps, *Nanoindentation.* Springer New York, 2004.

[91] Y. Sun, G. Lorimer, and N. Ridley, "Microstructure and its development in Cu-Al-Ni alloys," *Metallurgical and Materials Transactions A*, vol. 21, pp. 575 – 588, 1990.

[92] R. S. LiIlard and J. R. ScuIly, "Electrochemical passivation of ordered NiAl," *Journal of The Electrochemical Society*, vol. 145, pp. 2024 – 2032, 1998.

[93] J. Wharton and K. Stokes, "The influence of nickel-aluminium bronze microstructure and crevice solution on the initiation of crevice corrosion," *Electrochimica Acta*, vol. 53, no. 5, pp. 2463 – 2473, 2008.

[94] F. Çetinel, M. Müller, J. Rögner, W. Bauer, and J. Haußelt, *Influence of Dispersant on Rheology of Zirconia-Paraffin Feedstocks and Mechanical Properties of Micro Parts Fabricated via LPIM*, pp. 31 – 43. John Wiley & Sons, Inc., 2010.

[95] F. Çetinel, "Private communication prior to publication.".

[96] G. D. Quinn, *Fractography of ceramics and glasses, Special Publication 960-16.* NIST - National Institute of Standards and Technology, 2007.

[97] R. Dauskardt, W. Carter, D. Veirs, and R. Ritchie, "Transient subcritical crack-growth behavior in transformation-toughened ceramics," *Acta Metallurgica et Materialia*, vol. 38, no. 11, pp. 2327 – 2336, 1990.

[98] T. Kennerknecht, T. Straub, D. Buqezi-Ahmeti, and C. Eberl, "Very high cycle fatigue of micro moulded aluminium bronze using a custom built fatigue setup working up to 2 khz," in *VHCF5 Fifth International Conference on Very High Cycle Fatigue Proceedings* (C. Berger and H.-J. Christ, eds.), pp. 461 – 466, DVM, Berlin, June 2011.

[99] T. Kennerknecht, T. Straub, M. Funk, F. Çetinel, M. Härtelt, and C. Eberl, "Zuverlässigkeit von mikrourgeformten und nanostrukturierten Materialien unter Berücksichtigung der besonderen Anforderungen an Mikrobauteile," in *Kolloquium Mikroproduktion und Abschlusskolloquium SFB 499* (O. Kraft, A. Haug, F. Vollertsen, and S. Büttgenbach, eds.), pp. 75 – 82, KIT Scientific Publishing, Karlsruhe, October 2011.

[100] T. Straub, T. Kennerknecht, P. Robin, M. Tort, G. Kieffer, Y. Lapusta, and C. Eberl, "Small-scale multiaxial fatigue experiments in the very high cycle regime," in *VHCF5 Fifth International Conference on Very High Cycle Fatigue Proceedings* (C. Berger and H.-J. Christ, eds.), pp. 473 – 478, DVM, Berlin, June 2011.

Schriftenreihe des Instituts für Angewandte Materialien

ISSN 2192-9963

Die Bände sind unter www.ksp.kit.edu als PDF frei verfügbar oder als Druckausgabe bestellbar.

Band 1 Prachai Norajitra
Divertor Development for a Future Fusion Power Plant. 2011
ISBN 978-3-86644-738-7

Band 2 Jürgen Prokop
Entwicklung von Spritzgießsonderverfahren zur Herstellung von Mikrobauteilen durch galvanische Replikation. 2011
ISBN 978-3-86644-755-4

Band 3 Theo Fett
New contributions to R-curves and bridging stresses – Applications of weight functions. 2012
ISBN 978-3-86644-836-0

Band 4 Jérôme Acker
Einfluss des Alkali/Niob-Verhältnisses und der Kupferdotierung auf das Sinterverhalten, die Strukturbildung und die Mikrostruktur von bleifreier Piezokeramik $(K_{0,5}Na_{0,5})NbO_3$. 2012
ISBN 978-3-86644-867-4

Band 5 Holger Schwaab
Nichtlineare Modellierung von Ferroelektrika unter Berücksichtigung der elektrischen Leitfähigkeit. 2012
ISBN 978-3-86644-869-8

Band 6 Christian Dethloff
Modeling of Helium Bubble Nucleation and Growth in Neutron Irradiated RAFM Steels. 2012
ISBN 978-3-86644-901-5

Band 7 Jens Reiser
Duktilisierung von Wolfram. Synthese, Analyse und Charakterisierung von Wolframlaminaten aus Wolframfolie. 2012
ISBN 978-3-86644-902-2

Band 8 Andreas Sedlmayr
Experimental Investigations of Deformation Pathways in Nanowires. 2012
ISBN 978-3-86644-905-3

Band 9 Matthias Friedrich Funk
Microstructural stability of nanostructured fcc metals during cyclic deformation and fatigue. 2012
ISBN 978-3-86644-918-3

Band 10 Maximilian Schwenk
Entwicklung und Validierung eines numerischen Simulationsmodells zur Beschreibung der induktiven Ein- und Zweifrequenzrandschichthärtung am Beispiel von vergütetem 42CrMo4. 2012
ISBN 978-3-86644-929-9

Band 11 Matthias Merzkirch
Verformungs- und Schädigungsverhalten der verbundstranggepressten, federstahldrahtverstärkten Aluminiumlegierung EN AW-6082. 2012
ISBN 978-3-86644-933-6

Band 12 Thilo Hammers
Wärmebehandlung und Recken von verbundstranggepressten Luftfahrtprofilen. 2013
ISBN 978-3-86644-947-3

Band 13 Jochen Lohmiller
Investigation of deformation mechanisms in nanocrystalline metals and alloys by in situ synchrotron X-ray diffraction. 2013
ISBN 978-3-86644-962-6

Band 14 Simone Schreijäg
Microstructure and Mechanical Behavior of Deep Drawing DC04 Steel at Different Length Scales. 2013
ISBN 978-3-86644-967-1

Band 15 Zhiming Chen
Modelling the plastic deformation of iron. 2013
ISBN 978-3-86644-968-8

Band 16 Abdullah Fatih Çetinel
Oberflächendefektausheilung und Festigkeitssteigerung von niederdruckspritzgegossenen Mikrobiegebalken aus Zirkoniumdioxid. 2013
ISBN 978-3-86644-976-3

Band 17 Thomas Weber
Entwicklung und Optimierung von gradierten Wolfram/ EUROFER97-Verbindungen für Divertorkomponenten. 2013
ISBN 978-3-86644-993-0

Band 18 Melanie Senn
Optimale Prozessführung mit merkmalsbasierter Zustandsverfolgung. 2013
ISBN 978-3-7315-0004-9

Band 19 Christian Mennerich
Phase-field modeling of multi-domain evolution in ferromagnetic shape memory alloys and of polycrystalline thin film growth. 2013
ISBN 978-3-7315-0009-4

Band 20 Spyridon Korres
On-Line Topographic Measurements of Lubricated Metallic Sliding Surfaces. 2013
ISBN 978-3-7315-0017-9

Band 21 Abhik Narayan Choudhury
Quantitative phase-field model for phase transformations in multi-component alloys. 2013
ISBN 978-3-7315-0020-9

Band 22 Oliver Ulrich
Isothermes und thermisch-mechanisches Ermüdungsverhalten von Verbundwerkstoffen mit Durchdringungsgefüge (Preform-MMCs). 2013
ISBN 978-3-7315-0024-7

Band 23 Sofie Burger
High Cycle Fatigue of Al and Cu Thin Films by a Novel High-Throughput Method. 2013
ISBN 978-3-7315-0025-4

Band 24 Michael Teutsch
Entwicklung von elektrochemisch abgeschiedenem LIGA-Ni-Al für Hochtemperatur-MEMS-Anwendungen. 2013
ISBN 978-3-7315-0026-1

Band 25 Wolfgang Rheinheimer
Zur Grenzflächenanisotropie von $SrTiO_3$. 2013
ISBN 978-3-7315-0027-8

Band 26 Ying Chen
Deformation Behavior of Thin Metallic Wires under Tensile and Torsional Loadings. 2013
ISBN 978-3-7315-0049-0

Band 27 Sascha Haller
Gestaltfindung: Untersuchungen zur Kraftkegelmethode. 2013
ISBN 978-3-7315-0050-6

Band 28 Stefan Dietrich
Mechanisches Verhalten von GFK-PUR-Sandwichstrukturen unter quasistatischer und dynamischer Beanspruchung. 2013
ISBN 978-3-7315-0074-2

Band 29 Gunnar Picht
Einfluss der Korngröße auf ferroelektrische Eigenschaften dotierter $Pb(Zr_{1-x}Ti_x)O_3$ Materialien. 2013
ISBN 978-3-7315-0106-0

Band 30 Esther Held
Eigenspannungsanalyse an Schichtverbunden mittels inkrementeller Bohrlochmethode. 2013
ISBN 978-3-7315-0127-5

Band 31 Pei He
On the structure-property correlation and the evolution of Nanofeatures in 12-13.5% Cr oxide dispersion strengthened ferritic steels. 2014
ISBN 978-3-7315-0141-1

Band 32 Jan Hoffmann
Ferritische ODS-Stähle – Herstellung, Umformung und Strukturanalyse. 2014
ISBN 978-3-7315-0157-2

Band 33 Wiebke Sittel
Entwicklung und Optimierung des Diffusionsschweißens von ODS Legierungen. 2014
ISBN 978-3-7315-0182-4

Band 34 Osama Khalil
Isothermes Kurzzeitermüdungsverhalten der hoch-warmfesten Aluminium-Knetlegierung 2618A (AlCu2Mg1,5Ni). 2014
ISBN 978-3-7315-0208-1

Band 35 Magalie Huttin
Phase-field modeling of the influence of mechanical stresses on charging and discharging processes in lithium ion batteries. 2014
ISBN 978-3-7315-0213-5

Band 36 Christoph Hage
Grundlegende Aspekte des 2K-Metallpulverspritzgießens. 2014
ISBN 978-3-7315-0217-3

Band 37 Bartłomiej Albiński
Instrumentierte Eindringprüfung bei Hochtemperatur für die Charakterisierung bestrahlter Materialien. 2014
ISBN 978-3-7315-0221-0

Band 38 Tim Feser
Untersuchungen zum Einlaufverhalten binärer alpha-Messinglegierungen unter Ölschmierung in Abhängigkeit des Zinkgehaltes. 2014
ISBN 978-3-7315-0224-1

Band 39 Jörg Ettrich
Fluid Flow and Heat Transfer in Cellular Solids. 2014
ISBN 978-3-7315-0241-8

Band 40 Melanie Syha
Microstructure evolution in strontium titanate Investigated by means of grain growth simulations and x-ray diffraction contrast tomography experiments. 2014
ISBN 978-3-7315-0242-5

Band 41 Thomas Haas
Mechanische Zuverlässigkeit von gedruckten und gasförmig abgeschiedenen Schichten auf flexiblem Substrat. 2014
ISBN 978-3-7315-0250-0

Band 42 Aron Kneer
Numerische Untersuchung des Wärmeübertragungsverhaltens in unterschiedlichen porösen Medien. 2014
ISBN 978-3-7315-0252-4

Band 43 Manuel Feuchter
Investigations on Joule heating applications by multiphysical continuum simulations in nanoscale systems. 2014
ISBN 978-3-7315-0261-6

Band 44 Alexander Vondrous
Grain growth behavior and efficient large scale simulations of recrystallization with the phase-field method. 2014
ISBN 978-3-7315-0280-7

Band 45 Tobias Kennerknecht
Fatigue of Micro Molded Materials – Aluminum Bronze and Yttria Stabilized Zirconia. 2014
ISBN 978-3-7315-0293-7